IEE Telecommunications Series 12

Series Editors: Prof. J.E. Flood
C.J. Hughes

SPREAD SPECTRUM IN COMMUNICATION

Previous volumes in this series

Volume 1	Telecommunications networks J.E. Flood (Editor)
Volume 2	Principles of telecommunication-traffic engineering D. Bear
Volume 3	Programming electronic switching systems M.T. Hills and S. Kano
Volume 4	Digital transmission systems P. Bylanski and D.G.W. Ingram
Volume 5	Angle modulation: the theory of system assessment J.H. Roberts
Volume 6	Signalling in telecommunications networks S. Welch
Volume 7	Elements of telecommunications economics S.C. Littlechild
Volume 8	Software design for electronic switching systems S. Takemura, H. Kawashima, N. Nakajima Edited by M.T. Hills
Volume 9	Phase noise in signal sources W.P. Robins
Volume 10	Local telecommunications J.M. Griffiths (Editor)
Volume 11	Principles and practice of multi-frequency telegraphy J.D. Ralphs

R. Skaug
J. F. Hjelmstad

SPREAD SPECTRUM IN COMMUNICATION

Peter Peregrinus Ltd on behalf of The Institution of Electrical Engineers

Published by: Peter Peregrinus Ltd., London, UK.

British Library Cataloguing in Publication Data

Skaug, R.
Spread spectrum in communication.
— (IEE telecommunications series; v.12)
1. Communications, Military II. Spread spectrum communications
I. Title II. Hjelmstad, J.F. III. Series
355.8'5 UG590

ISBN 0-86341-034-0

Printed in England by Short Run Press Ltd., Exeter

Preface

Spread spectrum communications involve signalling techniques which greatly expand the transmitted spectrum relative to the data rate. That is to say, the bandwidth of the transmitted signals is much larger than the reciprocal of the duration of the fundamental signalling interval.

Although current applications for spread spectrum are primarily in military communications – a field familiar to the authors – there is a growing interest in this technique for mobile radio networks and for both communication and positioning applications in satellites.

This book focuses on spread spectrum communication systems using burst format message signalling. That is to say, message transfer may take place on a highly irregular or 'push-to-talk' basis and there is no information in advance on when to expect the information transfer to take place. Such systems must make use of reliable and fast acquisition and synchronization techniques. In particular, the development of portable and tactical spread spectrum radio links or networks with limited resources presents special challenges. Ease of management, fast establishment of communication, reliability, flexibility, low weight and low cost are all important factors influencing the feasibility and success of these spread spectrum systems.

The means by which the spectrum is spread is crucial. The main portion of the theory developed in this book deals with direct sequence spread spectrum systems. The requirements for fast and reliable synchronization of radio transmitters and receivers means that signal acquisition and the correlation process should be performed by matched filters.

The primary motivation for this book has been provided by our association with spread spectrum communication development at the Norwegian Defence Research Establishment (NDRE) and the environmental surveillance technology programme of the Royal Norwegian Council for Scientific and Industrial Research. The material in this book has been gathered from many sources – technical papers, internal reports, private conversations, contributions from our colleagues and our personal experiences. The concept of generalized spread spectrum coding schemes was introduced by Professor Dag T. Gjessing, and the elaborations on this theme found in Chapter 2 are based on his suggestions.

The number of spread spectrum topics is vast, and this is reflected in the numerous technical articles on the subject. To adequately cover spread spectrum communication systems in general, several volumes would be required. We have however attempted to limit our presentation to results and concepts with which we have become familiar during our research work.

The authors have benefited from numerous discussions with their colleagues, who have also supplied helpful material and figures. In particular the authors wish to acknowledge the contributions to Chapter 4 on LSI and SAW technology of O. Søråsen, NDRE and K. A. Ingebrigtsen, Electronics Research Laboratory, The Norwegian Institute of Technology. The authors have also been assisted by valuable notes, relating to the many aspects of spread spectrum modulation, by T. Thorvaldsen, N. A. Sæthermoen, K. Besserudhagen, H. Olaisen and I. Koltveit of NDRE and Å. T. Johansen, Research Establishment, Norwegian Telecom Administration. A note of appreciation also goes to N. Klippenberg and M. Norland who made available results of their diploma work at the Norwegian University of Technology, Trondheim on narrowband interference cancellation. A personal note of gratitude from one of us (R. Skaug) goes to S. Heier for his long participation in software development for code analysis, and to D. Brodtkorb, promoter of spread spectrum activities for several years at NDRE. We would like to express our sincere appreciation to the Electronics Division, NDRE and to the head of the division, N. Holme, who has provided extensive facilities for the writing of this book.

Finally we would like to give special thanks to the magician behind the manuscript, Miss H. Stensrud, who works the word processor much after the principle of the water in a Norwegian waterfall. She gives an intrepid glance at the top of the page; the text then roars to the bottom, after which the finished sheet silently leaves the machine. We would also very much like to acknowledge the valuable services rendered by the drawing-office, and in particular those of Miss A. Nymoen and Mrs U. Næss.

Contents

List of symbols

A	signal amplitude
$A(f)$	amplitude spectrum
$a(t)$	spreading code
α^{r}	power ratio between received signal in a CDMA network
B	data bandwidth
$b(t)$	data signal
$C(l)$	aperiodic correlation
$C(t, f)$	time–frequency correlation function
C_i	coefficients for a feedback shift register
d	antenna separation
D	Doppler (hertz)
E	energy per data bit
$\boldsymbol{E}(f)$	electric field vector
$E\|$	expected value of variable
E_{space}	efficiency of space domain (antenna) operations
E_{system}	total efficiency of electronic measures
E_{time}	efficiency of time domain operations
ϕ	Gaussian cumulative distribution
$F(a)$	merit factor for binary sequences
$f(\tau, t - s)$	fading process
$f(x)$	characteristic polynomial
$f(x)$	minimal polynomial
γ_E	error coding improvement factor
$h(t)$	filter impulse response
$H(\chi)$	entropy of symbol χ
$\hat{\boldsymbol{h}}_H$	horizontal unit vector
η^R	antenna efficiency for reception
η^T	antenna efficiency for transmission
$\hat{\boldsymbol{h}}_v$	vertical unit vector
I	interference
$j(t)$	jammer field strength
K	number of users in a multiple access network

L	range delay (seconds)
m	number of stages in feedback shift register
$n(t)$	noise process
N_0	noise spectral density
ω_c	carrier frequency
p	period of a spreading code
P	transmitted power
$p(\chi)$	probability of symbol χ
P_d	probability of detection
P_e	probability of error
P_{fa}	probability of false alarm
PG	processing gain value
p_T	rectangular pulse function
Q	channel delay cross power spectral density
θ	even correlation for codes
$\hat{\theta}$	odd correlation for codes
$q(a)$	decimation of sequence
Q^{kr}	code interference
ϱ	angle to interference
$R(t, s, \tau, z)$	channel correlation function
$R(\tau)$	autocorrelation function
$r(t)$	received signal
R_{AC}	distance from A to C
R_c	spreading code chip rate
R_d	data rate
ϱ_f	normalized fading envelope covariance
S	desired signal component
$S(a^k)$	code sidelobe energy
$S(f)$	power spectral density of sequence
$\sigma(r, f)$	channel scattering function
$s(t)$	signal waveform
$S_D(f)$	Doppler scatter function
σ_f^2	fading envelope variance
S^i	ith moment of spreading sequence
S^k	sidelobe energy for a spreading code
$S_r(\lambda)$	range scatter function
S_t	state in a recurrence relation
t	bound on crosscorrelation due to Gold
T	correlation time
t_c	fraction of a spreading code chip
T_c	spreading code chip duration
T_d	data bit duration
t_s	sampling rate
$T^s(a)$	cyclic permutation of code

V_I	interference signal voltage
V_R	voltage at receive antenna terminal
V_T	voltage at transmit antenna terminal
W	spread spectrum bandwidth
$w(t)$	window function
$Y(t)$	correlator output

Chapter 1

Introduction to spread spectrum communications

1.1 Intentions of the book

Spread spectrum techniques represent a fairly recent contribution to the communications community. There has been rapid and continuous progress in communication technology and theory ever since information was first carried by electromagnetic waves at the end of the nineteenth century. Technology and theory have always been closely linked, no more so that in spread spectrum communications. Advances in technology have enabled the radio engineer to utilize previously unforeseen processing capacity in communication networks, and high performance computers have provided theoreticians with tools required for performance analysis and modelling.

This book offers an introduction to the methods and concepts applied in modern communications, and suggests areas for future development and refinement. The term *spread spectrum* is used to describe a class of modulation techniques in which the data stream is coded in such a way that the total information rate is higher than the data rate. However, this book also describes a range of other electronic measures which are available to the engineer. It is therefore necessary to introduce the degrees of freedom given by the physical nature of electromagnetic waves and to give an overview of the corresponding possibilities in communication engineering. In order to design optimal communication systems, it is of vital importance that the essentials of the problems and possibilities of the various communication techniques are thoroughly understood.

As well as unified descriptions of the various communication techniques, a basis for mathematical analysis is included. Because detailed mathematical treatment requires specific information about the system, an introductory analysis only is presented together with examples and applications. This will enable the radio engineer to insert the actual parameters and develop the equations for his particular application.

1.2 Contents of the book

To enable a distinction to be made between a spread spectrum communication system and a conventional radio system, an introduction to the basis of spread spectrum communications and the associated terminology is given in Chapter 1. It is assumed that the reader is familiar with radio engineering and basic signal analysis. The relationship between spread spectrum modulation and conventional RF modulation is discussed and so is the relative merit of spread spectrum modulation and encryption for data security. The choice of spread spectrum modulation also influences the requirement for error coding and source coding. This is due to the increased resistance to interference provided by spread spectrum modulation. Furthermore, the matched filter concept for signal acquisition in a spread spectrum communication system is introduced.

The requirement for synchronization of the time references in the transmitter and receiver leads to a split mode operation of the modem, in which time synchronization and data transmission are performed successively. Depending upon the characteristics of the system and interference, the emphasis put on the synchronization and the data transmission will be different.

There is also some confusion about the interpretation of the term *processing gain*. The processing gain is a measure of the quality of a spread spectrum system, and a high processing gain corresponds to high interference rejection and low error rate. An analytic formulation of the processing gain for spread spectrum systems is derived, and the interference rejection capability for a passive matched filter is compared with various interpretations of the processing gain of the system.

Chapter 2 extends the definition of spread spectrum communications to include the full range of principles and techniques which can be applied to achieve the ultimate communication network. The various properties of electromagnetic waves are exploited to allow the intelligent user to optimize his communication system with respect to each property. Specifically, the spatial distribution of field strength may be utilized for adaptive antennas for interference rejection and signal maximization, and the polarization properties of the electromagnetic wave can be very useful for coding and signal/noise maximization.

In communications there is always the possibility of interference both from natural sources and from other radio links. Usually interference from the latter is kept as low as possible through frequency allocation agreements and by adequate spacing of stations that may interfere. In military communications, hostile actions will take place. Intentional jamming and eavesdropping can be anticipated and, although these are basically the same phenomena as in a nonmilitary system, there are a few characteristics peculiar to military communications. First of all, jamming and eavesdropping will be used as part of some deliberate military action. Secondly, the situation is dynamic in that each improvement will cause actions by the other participant in the battle.

To be able to analyse the behaviour of a dynamic military communication network, electronic warfare scenario modelling is needed. The models are described by operations analysis equations, and permit prediction of success or failure of specific developments in electronic warfare scenarios. This approach for describing the effect of mutual disturbance is also useful for civilian applications.

A systematic approach to the various electronic warfare techniques is then introduced. By analysing a communication system as a set of independent subsystems, a simplified treatment of the various electronic warfare techniques is achieved. Time and space coding techniques are examples of independent operations. Intelligence is also discussed; the consequences are considered of some knowledge of system structure or system code becoming known to the opposing party.

To simplify the presentation and analysis of time domain coding schemes, frequency domain representation of the time series is given using time–frequency maps. These maps are used to describe all the various spreading techniques. This unified description of frequency hopping, time hopping and direct sequence techniques facilitates a generalized approach to the analysis of crossinterference and jamming in an environment with a multitude of systems.

The electronic countermeasures to the various systems are then analysed. To be able to distinguish between actions in various environments, three types of environments are considered. The first assumes that each participant in the electronic warfare game is totally ignorant about the dispositions of the other participants. This is the classical military assumption. The success of various strategies for eavesdropping, detection, localization and jamming is then calculated and an easily visualized technique for success analysis is presented.

A more common situation is when some of the methods and equipment of the other participants are known. This can be the case when intelligence data are available; moreover, monitoring and electronic surveillance can provide information which can be utilized for electronic countermeasures.

The third situation is when each communication network operates independently, but full knowledge of all the systems parameters and codes are available to every participant. This is the classical civilian communication scenario, and this chapter provides an approach to the understanding of crossinterference as well as presenting simple mathematical tools for calculating the impairment in performance. This third case can also be considered for military scenarios when full knowledge of the systems involved exists. For example, when old systems are used it can be assumed that all participants will have access to the technology. Also, it is well known that, with time, intelligence actions and even unclassified information flow prevent the maintenance of full security of a military system.

Space domain coding is then discussed. The obvious applications are adaptive antennas with null steering and polarization agility. These are very important aspects, as these techniques are independent of whatever coding might be applied in the time domain. The success of the various techniques is analytically presented in the space domain success functions.

This implies that the total performance of a system utilizing a set of time and space domain coding techniques can be found by combining all the success functions. This is illustrated in an example where a set of conventional and spread spectrum radio links operate in the same area.

The generation and properties of spreading codes for direct sequence systems are the subject of Chapter 3. The problems associated with synchronization, multipath and spread spectrum multiple access, allowing many users to transmit simultaneously in time and occupy the same RF bandwidth, require a careful selection of spreading codes based on their different correlation properties. The type of code used sets bounds on the capability of a system, bounds that can only be changed by changing the codes. Maximal length sequences and Gold sequences are probably the most frequently used spreading sequences today. The chapter develops key performance parameters and evaluates the advantages and shortcomings of these codes. Large tables with important code performance parameters are given, enabling engineers to delimit system performance. Another important aspect of spread spectrum modulation, that of signal concealment, is also shown to be related to the choice of spreading codes. Finally, the relation between spreading codes and the required hardware processing complexity is briefly treated.

An extremely important aspect of advanced communication systems is the matching of hardware electronic technology and signal processing algorithms, and this is treated in Chapter 4. The structure of the communication terminal into its functional blocks is presented, and the hierarchy of processing levels performing signal decoding, synchronization, signal flow control and network control is described.

Signal processing technology implies that the best possible algorithm should be used to obtain full benefit from the best component available. To do this, component manufacturers should consider making components that can perform complex tasks as well as simple arithmetic and functional operations. Today, it is feasible to construct devices that utilize physical phenomena to perform even very complex mathematical operations. These components have a spatial structure that enables a large number of operations to be performed in parallel. Examples include acousto-optic Fourier transformers and surface acoustic wave convolvers.

Recent designs of spread spectrum communication systems are reported; special emphasis is put on direct sequence spread spectrum communication modems. The choice between digital and analogue processing is an important issue in spread spectrum systems. There is no golden rule as to which is better for a specific application, and sound engineering judgement has to be used in each case.

Chapter 5 introduces the propagation medium, which cannot be ignored in practical system considerations. A characterization of the channels important to spread spectrum modulation is developed to aid in the prediction of system results. Issues of particular concern relating to certain spread spectrum communication channels include available 'coherent' bandwidths and possible signalling speeds. The related issue of time variant multipath and system difficulties in keeping transmitters and receivers synchronized is also discussed. A propagation measurement program and its results are reported for high frequency ionospheric channels. Channel simulation and computer prediction models can be important tools for indicating such fundamental aspects of communication as user capacities and cochannel interference. Statistical estimates of the success of a spread spectrum multiple access network under fading conditions are presented.

Chapter 6 is devoted to direct sequence spread spectrum code division multiple access networks. Both the influence of spreading code interference and the problem of operating spread spectrum systems in the presence of interference from conventional radios are considered. Since spread spectrum systems make use of large portions of a radio frequency band it is unlikely that these systems will be the sole users within this band. In particular, Chapter 6 gives examples of possible methods for narrowband interference suppression. The suppression may take place either by frequency filtering or by spatial filtering, the latter usually being referred to as antenna null steering.

1.3 Spread spectrum modulation compared with conventional methods

To introduce spread spectrum modulation in terms familiar to communication and radio engineers, a communication link divided into its functional blocks is presented in Fig. 1.1. The subunits are characterized by their function, and the particular configuration of functional blocks has been fixed during the development of steadily more sophisticated and advanced systems. Each improvement has added complexity to the communication system and required new theoretical methods and mathematical tools. In addition, it will be found that for various systems the relative importance of each of the functional blocks will be different.

1.3.1 *Source coding*

The input data stream is assumed to consist of a given set of symbols. The objective of the source coding is to convert the original symbol space to a symbol space which requires the minimum channel capacity. This is obtained when the length of each symbol is made proportional to its relative probability of occurrence. An alternative formulation is to require the input data stream to have maximum entropy.

The symbol set is denoted by $X = (x_1, x_2, \ldots, x_N)$ and the relative probability of occurrence of each symbol by $p(x_i)$. It follows that

$$\sum_{i=1}^{N} p(x_i) = 1 \tag{1.1}$$

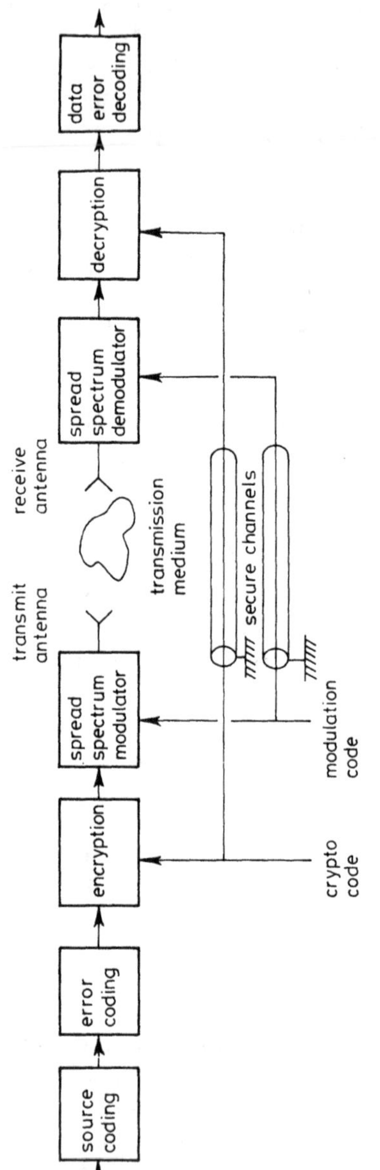

Fig. 1.1 *Functional blocks of a communication link*

It is obvious that more information is received in the detection of a symbol with a low probability of existence than in the detection of a more probable symbol. The total mean information content (entropy) $H(x)$ of all symbols is

$$H(x) = \sum_{i=1}^{N} p(x_i)I(x_i) \tag{1.2}$$

where $I(x_i)$ is the information content of each symbol x_i. As the requirement is that the information rate should be maximum when all symbols are equiprobable,

$$p(x_i) = 1/N \tag{1.3}$$

then the relationship between $p(x_i)$ and $I(x_i)$ should be a logarithmic one of the form

$$I(x_i) = \log_B \frac{1}{p(x_i)} \tag{1.4}$$

where B is the base of the logarithmic function.

This gives the following for the total mean information:

$$H(x) = \sum_{i=1}^{N} p(x_i) \log_B [p(x_i)]^{-1} \tag{1.5}$$

Usually the base chosen is $B = 2$. By convention, the corresponding unit for the information content is bits/symbol. If the decimal logarithmic function is used, the unit is decit/symbol (or hartley/symbol), and the entropy measured by the natural logarithmic function is given as nats/symbol.

1.3.2 *Error coding*

The purpose of error coding is to trade information rate capacity with information error rejection. This is achieved by adding to the transmitted symbols a set of symbols which do not carry information, but which are derived from the nonredundant set of symbols in such a way that they can detect and also correct errors in the information carrying symbols.

The symbol list is thus

$$X = (x_1, x_2, \ldots, x_N, y_1, y_2, \ldots, y_M) \tag{1.6}$$

where y_1 to y_M are the M redundant symbols with probabilities $p(y_i)$.

The corresponding reduction in channel capacity relative to the non error coded channel is given by

$$\sum_{i=1}^{M} p(y_i) = 1 - \sum_{i=1}^{N} p(x_i) \tag{1.7}$$

The upper bound for the improvement in error performance will be given by the relative distribution of energy in the redundant and nonredundant symbols:

$$\gamma_E = \frac{\sum_{i=1}^{M} p(x_i)}{\sum_{i=1}^{N} p(x_i)} = \frac{\sum_{i=1}^{N} p(y_i)}{1 - \sum_{i=1}^{M} p(y_i)} = \frac{1}{\sum_{i=1}^{N} p(x_i)} - 1 \tag{1.8}$$

For an actual error coding scheme, the improvement factor will be lower than the optimal figure γ_E.

1.3.3 *Encryption*

To deny unauthorized access to the communication system, the data stream can be converted into another data stream in such a way that the time or cost associated with the decoding of these scrambled data will be very much higher for the nonauthorized receiver than for the intended receiver. Also the authentication properties of encryption are important in guarding against spurious data being introduced into the receiver.

If the input data stream is given by $b(t)$, the encryption unit produces a new code $e(t)$ by an algorithm $F(\ )$, where

$$e(t) = F[b(t)] \tag{1.9}$$

The decryption unit reproduces the original data flow by the application of the inverse transform to the new code:

$$b(t) = F^{-1}[e(t)] \tag{1.10}$$

The unauthorized user will have to use algorithm $G(\ )$, which is a substitute for the $F(\ )$ algorithm when the code is not known.

The cost effectiveness of the encryption with respect to unauthorized reception is given by the ratio of the cost of performing the $G^{-1}(F)$ algorithm to that of the $F^{-1}(F)$ algorithm:

$$\text{encryption security} \sim \frac{\text{cost of } G^{-1}(F)}{\text{cost of } F^{-1}(F)}$$

The authentication security is given by the ratio of the costs of $F^{-1}(G)$ and $F^{-1}(F)$:

$$\text{authentication security} \sim \frac{\text{cost of } F^{-1}(G)}{\text{cost of } F^{-1}(F)}$$

It should be noted that the cost of the G and G^{-1} functions will depend dramatically on the data flow through the insecure channel (transmission medium) relative to that through the secure channel. If for every symbol transmitted there exists one or more code symbols, the transmission is totally secure. However, for most radio applications the code length will be shorter than the information stream length, and possible G functions may exist.

For multiuser systems or systems where no secure channels exist, the key can be presupplied to the receiver. The structure of the system can also be such that part of the key is presupplied to the receiver and part transmitted with the coded data.

1.3.4 *Spread spectrum modulation*

To facilitate the transfer of data through the transmission medium, the data stream has to be converted to a suitable radio frequency waveform. The traditional approach has been to modulate the data stream into a radio frequency carrier and stack the individual transmissions side by side in the frequency band. Frequency, amplitude or phase modulation is used, depending on the particular system design.

The importance of spread spectrum modulation techniques is that, by a substitution of the basic RF modulator by a modulator capable of generating waveforms with an information rate orders of magnitude higher than that of the data flow, a dramatic increase in transmission properties and system capability is obtained.

In a spread spectrum system, the basic idea is to replace a single data bit with a coded waveform. This data bit can then be detected optimally only when this code is known to the receiver. The more complicated and well designed is the code, the better and more secure will be the transmission.

Table 1.1 illustrates how spread spectrum modulation techniques compare with the other functions in a communication link. From this table it is seen that some operations contribute to a specific task in parallel with other functions, whereas other operations are not affected by processing elsewhere in the system. It is therefore obvious that a choice must be made in the allocation of processing power to achieve the required total operational characteristics throughout the system.

Table 1.1 shows that the basic data source optimization and error correction schemes are independent of the encryption and modulation processing that follow. Also, encryption and conventional RF modulation are orthogonal operations, whereas encryption and spread spectrum modulation may to a certain extent achieve the same goals. Spread spectrum modulation is a substitute for conventional RF modulation, and in addition has several operational advantages.

The relationship between encryption and spread spectrum modulation is extremely important. Whereas encryption is performed on the data bit stream itself, spread spectrum modulation operates on the RF waveform. Conventional encryption achieves eavesdropping and authentication security regardless of the state of other system variables such as the near–far phenomena, which will influence the performance of a spread spectrum system in this respect.

There are some very desirable features of spread spectrum modulation techniques. First of all, new schemes of band allocation can be devised. Code multiplexing can be utilized to stack a number of channels within the same frequency band; this implies that a set of data bits can be transmitted at the same time and in the same frequency band. Furthermore, unauthorized listening is

Table 1.1 *Spread spectrum modulation compared with conventional methods in communication*

	Source coding	Error coding	Encryption	RF modulation	Spread spectrum modulation
Data source optimization	*				
Error detection		*			
Error correction		*			*
Transmission security unregarded channel coding			*		
Band allocation				*	*
Protection against unauthorized reception (eavesdropping)			*		*
Protection against unauthorized transmission (authentication)			*		*
Protection against interference (jamming)					*
Protection against detection of radiation					*
Data error reduction for a given spectral density power					*
Optimization to transmission medium (multipath fading)					*

discouraged by making it impossible to receive the signal without knowing the code. This means that only a receiver matched to the code is useful. Authentication is also covered, as only the transmitter knowing the code will be able to generate the code which will pass into the receiver that is matched for the same code. Protection against interference is carried out in a similar way, as the interference will usually be correlated with the code only to a very limited extent. As a consequence, the matched receiver will not respond to the interference. When an interference signal has some characteristic properties, these can be exploited by making the receiver nonsensitive to this specific interference by using the proper code and modulation.

A spread spectrum transmission will in general be very hard to detect. This is because the probing receiver will not be matched and will therefore be very insensitive to coded transmissions. As a consequence, for a system which in a specific environment is only allowed to radiate a certain maximum spectral density, the total power transmitted can be increased by virtue of the fact that the coded waveform will be spread in the frequency domain. This means that the total received power per data bit and the overall system signal/noise ratio can be increased without the transmission interfering with other links or becoming more vulnerable to detection.

One very important aspect of spread spectrum modulation is that the generalized modulation scheme opens new avenues for optimal coding to overcome the negative effects of the transmission medium. Phenomena like multipath fading, time variable attenuation and transmission delay fluctuations are counteracted by the redundancy of the transmitted signal. In its ultimate mode, the communication link can also operate adaptively. If the properties of the transmission medium are investigated before transmission, the optimum code may be chosen.

1.4 Matched filters for spread spectrum modulation

As pointed out in Section 1.3, spread spectrum modulation implies that the data stream is substituted by a complex waveform which is decoded in a matched receiver. This receiver will at any time give an output signal which is proportional to the 'match' of the input signal to the expanded coded waveform. This 'match' is measured in terms of correlation at various instants of time.

From linear filter theory it is known that the optimum filter for detecting a time series $a(t)$ is one with a transfer function $H(f)$ equal to the complex conjugate of the amplitude spectrum of $a(t)$. This gives the minimum signal error for white, additive Gaussian noise.

Thus the receiver will have to perform an operation of the form

$$\Sigma a(t)h(\tau - t)\,\mathrm{d}t \qquad (1.11)$$

If the code length is M chips, then for each value of τ there are required M multiplications and $M - 1$ additions. For a nonsynchronized system the time lag τ is not known, and this process has to be repeated until match is obtained. For the consecutive data bits, however, the time lag is known and the correlation process will have to be performed only at the instants of time when the filter is matched.

The method considered for initial synchronization in burst communication systems is based on passive correlation utilizing matched filters. The output of the matched filter shows an impulse-like correlation peak at the instants of time at which the incoming spreading code corresponds to the spreading code programmed into the matched filter. These instants are p/R_c apart and acquisition is in principle completed by the first detected correlation peak. Ideally, then, when the system is not corrupted by noise the mean acquisition time is $T_{acq} = p/2R_c$. In reality this estimate is far too simple to be of any use when calculating the realizable synchronization time for burst message signalling. The signal will obviously be corrupted by noise and interference, and it is as important to realize that an appropriate signal/noise ratio for data demodulation may be far from acceptable for initially acquiring the correlation peak with satisfactory probabilities of detection and false alarms. When data detection takes place a time window is inserted around the expected time of arrival for the correlation peak, whereas during initial synchronization there may be no *a priori* information on the time of arrival of the message.

Thus the signalling format requires a synchronization preamble which will allow the receiver to reliably detect that synchronization has started and to determine the exact time of arrival of the correlation peaks. This time information is then used to insert the required time window during data demodulation. A system not inserting such a time window will perform similarly to, and just as badly as, an active correlation system without a final filter matched to the data rate.

1.4.1 *Choice of detector for acquisition and synchronization*

The error rate P_e during data demodulation is measured as a function of the mean signal/noise power ratio $\gamma = [Aa(t)]^2/2N$, where $a(t)$ is an arbitrary spreading code of p chips each having a value ± 1, and A includes the path attenuation assumed constant over the transmission interval. The peak signal/mean noise ratio $[Aa(t)]^2/N = 2\gamma$ is for matched filters corrupted by white noise $2E/N_0$, where N_0 is the noise spectral density [1]. E is the signal energy, which for the bandpass and lowpass realizations takes the form

$$\begin{aligned} E_{\text{bandpass}} &= \int_0^{T_s} Aa(t)\cos[\omega t + \theta(t)]^2\,dt \\ &= \frac{A^2T_s}{2} = \frac{A^2pT_c}{2} \end{aligned} \tag{1.12}$$

$$\begin{aligned} E_{\text{lowpass}} &= \int_0^{T_s} \left\{\left[\frac{A}{2}a(t)\cos\eta\right]^2 + \left[\frac{A}{2}a(t)\sin\eta\right]^2\right\}dt \\ &= \frac{A^2T_s}{2} = \frac{A^2pT_c}{2} \end{aligned} \tag{1.13}$$

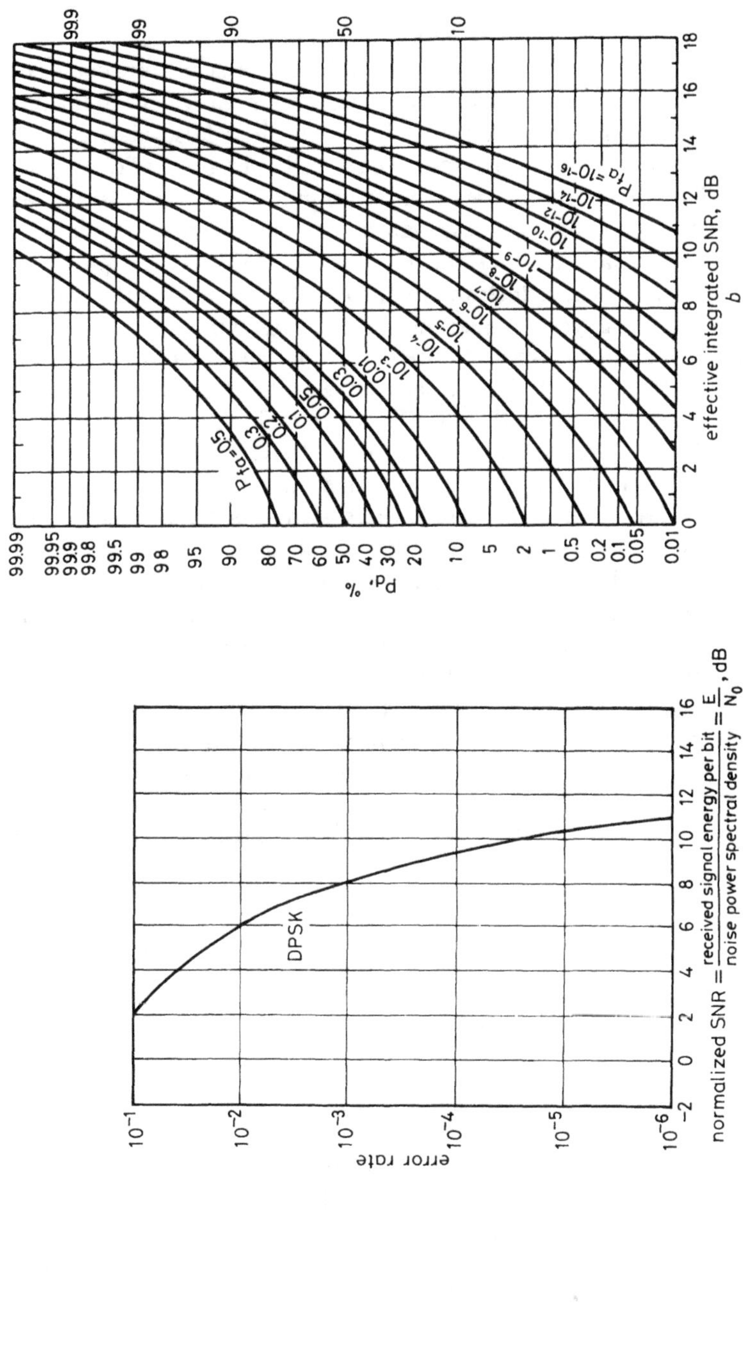

Fig. 1.2 *a* Error rate for DPSK demodulation
b Probability of detection of correlataion pulses in Gaussian noise

For a specified type of modulation the necessary signal/noise ratio can be found for the presupposed error rate. Now this signal/noise ratio should as far as possible match the requirements for probability of detection and false alarms during the time allowed for synchronization. The problem of synchronizing to a correlation peak is similar to the detection problem is pulsed radar systems. If for simplicity the output is supposed to be sampled at the spreading code rate, the number of cells in which to look for synchronization is equivalent to the data bit/spreading chip ratio PG (see Section 1.5). If the final false alarm rate is required to be P_{fa}, then the false alarm rate in each cell must be P_{fa}/PG. In order to reliably establish the point of peak correlation to be used during data detection, the first correlation peak estimate is sometimes checked against a number of consecutive peaks at times one data bit apart. Each correlation peak then requires a probability of detection equal to $P_d^{1/M}$, where M is the number of peaks used to establish synchronization and the time window.

The problem with a simple threshold detector now becomes apparent. The probability of undetected correlation peaks will be quite large for medium and low signal/noise ratios because the threshold must be set high if a low false alarm rate is required. A low false alarm rate is usually desired because each alarm requires time before it can be rejected. On the other hand the possibly high false rejection will also increase the time before the system has acquired synchronization. Fig. 1.2 shows that, in addition to these problems, an acceptable probability of detection and false alarm during synchronization for a simple threshold detector will usually require a much higher signal/noise ratio then that required to achieve reasonable error rates during data detection. As an example, a signal/noise ratio of 13 dB gives a probability for detecting a correlation peak of 90% at a false alarm rate of 10^{-6}, where for DPSK data demodulation the same signal/noise ratio gives an error rate before error coding of $\approx 10^{-9}$, which is unnecessarily low for most purposes. Similarly, a signal/noise ratio of 8 dB, giving an error rate of 10^{-3} during data demodulation, will only give a probability of detection of 5% for a false alarm rate of 10^{-6} during synchronization.

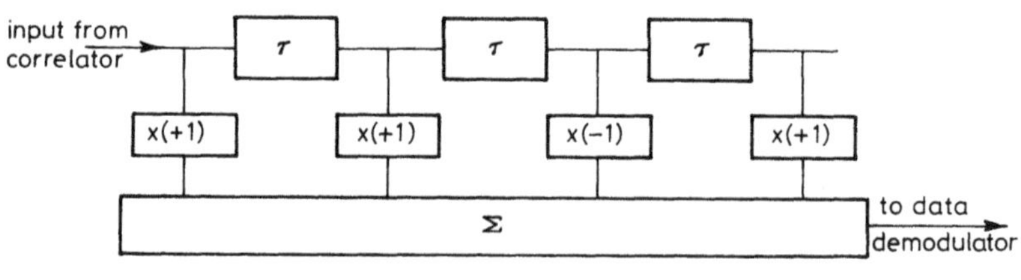

Fig. 1.3 *Transversal filter for integration of correlation pulses*

Somehow a better match between performance during data demodulation and synchronization must be achieved. A possible method to increase the probability of detection without increasing the false alarm rate is to make use of a transversal filter that effectively integrates the correlation pulses, as shown in Fig. 1.3. If the correlation pulses in the preamble are coded as a Barker code, an impulse-like correlation peak will appear at the output when the preamble completely fills the filter. The achievement possible when integrating correlation pulses is shown in Fig. 1.4. For example, the filter will, for a signal/noise ratio of 8 dB, increase the probability of detection from 5% to 90% at a constant false alarm rate of 10^{-6} if four correlation pulses are integrated.

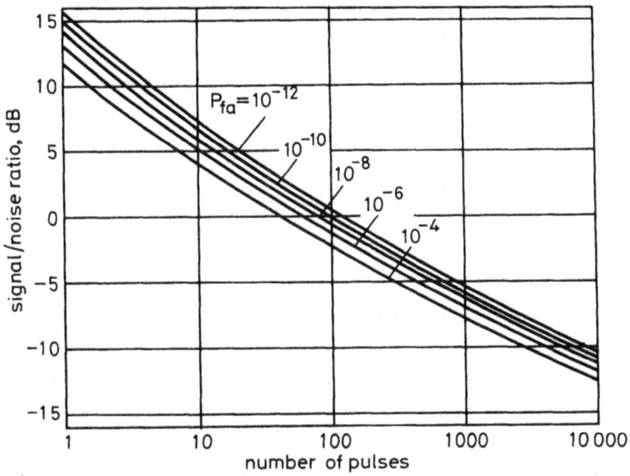

Fig. 1.4 *Required signal/noise ratio as a function of the number of correlation pulses integrated for a linear detector. Probability of detection* =0.9
(BLAKE, L. V., 'A guide to basic pulse-radar max range calc', NRL-Report 5868, December 62.
By courtesy of NRL (Department of the Navy, Naval Research Laboratory, U.S.A.))

The delays necessary in the transversal filter can be produced in either digital or analogue form. A digital solution using a feedback integrator with the delays implemented by shift registers is shown in Fig. 1.5. The circuit is composed of an A/D converter, a summation unit, a shift register and finally a multiplication unit. The length of the shift register matches the length of a data bit so that the correlation peaks are added in phase. The multiplication unit with a factor $K < 1$ is inserted to make sure that the sum does not grow infinitely, since the

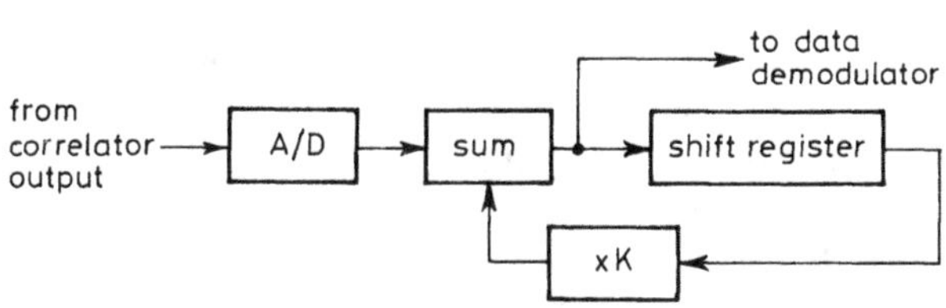

Fig. 1.5 *Integrator circuit based on digital delay line implemented with shift registers*

time of arrival of the synchronization preamble is not known and is thus not a starting point for the integration. The performance for such a circuit will be poorer than the theoretical curves (where K can be set to 1) because of time of arrival information. As an example, signal/noise ratio of 8 dB and $P_d = 90\%$ and $P_{fa} = 10^{-6}$ using a transversal filter integrating four correlation pulses is used to compare results. Simulation shows that the feedback integration achieves $P_d = 90\%$ for $P_{fa} = 10^{-6}$ when the signal/noise ratio is ≈ 10.5 dB with $K = 0.7$. Compared with an ideal integration the loss is 2.5 dB. In order to reduce the required signal/noise ratio, additional pulses must be integrated.

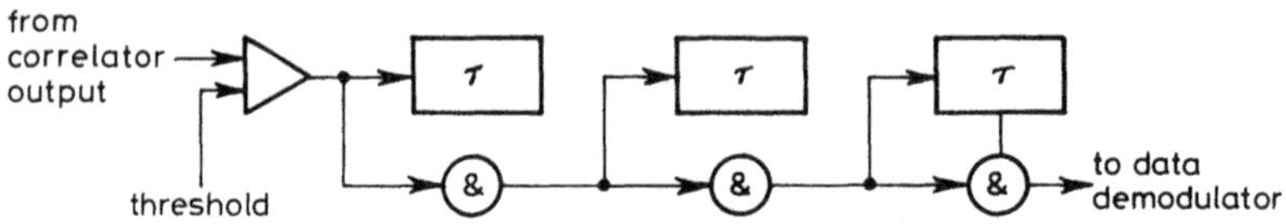

Fig. 1.6 *Coincidence detector*

Another method to improve synchronization performance is to lower the false alarm rate using coincidence detection. A suitable circuit is shown in Fig. 1.6. This circuit does not involve the dynamics necessary for the shift registers in the integrator system. First the output from the passive correlator is compared with a threshold. The output from the comparator is then moved into a shift register. The output from each shift register stage is fed through an AND gate to the next stage as shown. If for example a 1 from the comparator coincides with a 1 from the first register stage, a 1 is fed to the next register stage, and so on. Thus correlation pulses one shift register stage apart will result in all 1 stream at the output after $M + 1$ clock pulses. Let P_{di} and P_{fai} be the probabilities of detection and false alarm at the input of the circuit and let P_{do} and P_{fao} be the probabilities of detection and false alarm at the output of the circuit. Then

$$P_{do} = (P_{di})^{M+1} \tag{1.14}$$

$$P_{fao} = (P_{fai})^{M+1} \tag{1.15}$$

It is now possible to lower the threshold in the comparator to a value giving a sufficient probability of detection, since the resulting higher false alarm rate is lowered by the requirement for $M + 1$ consecutive false alarms at the input to give one false alarm at the output.

The pulse(s) at the output of the synchronization detector will only allow the system to decide that a synchronization preamble is being received. To accurately decide the time for the correlation peaks requires additional hardware. Several methods to perform this task are possible and usually involve a digital or analogue peak detector. However, a somewhat different method will be outlined here. Consider the coincidence detector of Fig. 1.6 with an extra shift register clocked at a higher rate than the coincidence register. The circuit is shown in Fig. 1.7 together with two possible phases of the clocks relative to a correlation peak. In the left-hand timing diagram the coincidence register is clocked just after the correlation peak exceeds the threshold in the comparator.

The resulting time diagram at the input of the AND gate is shown for the case of no false alarms and three stages in the extra register. Three occasions of false alarms are however possible (encircled), influencing the extra register. The occurrence of any of these alarms will give a nonoptimum sampling point affecting the data error rate. For the situation shown in the right-hand timing diagram the correlation peak is sampled by the coincidence detector just before the correlation peak exceeds the threshold, and no incorrect sampling point is possible.

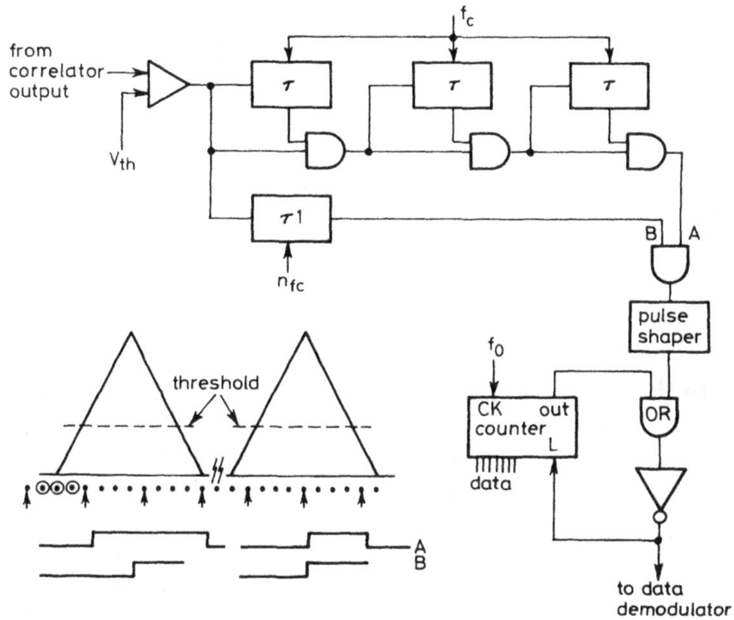

Fig. 1.7 *Complete circuit for signal acquisition and data demodulator sampling using a coincidence detector*

Whether or not it is possible to use the sampling point obtained during the synchronization period throughout data demodulation will depend on message length and clock stability. If sufficient accuracy is not achievable, the sampling point has to be re-established, or in practice the sampling outputs from the synchronization unit must be used throughout the data demodulation. A problem does however arise in this situation because $P_d \neq 1$ for the synchronization circuit. It therefore becomes necessary to prevent loss of data bits due to loss of detection in the synchronization circuit. The coincidence synchronization unit will for example give a miss of $M + 1$ data bits if one correlation peak goes undetected in the synchronization circuit. Additional hardware is necessary to avoid this situation. A counter can be inserted as shown in Fig. 1.7. The counter could for examle have a synchronous input and a 'ripple carry' output which goes high when the counter contains all 1s. When the input L is put low the contents of the data inputs are loaded into the counter on the first rising clock pulse. The data inputs are set such that the counter reaches the all 1 count during the period between one correlation peak and the next. Connecting an inverted 'ripple carry' output back to the L input will produce an output at time instants exactly matching the distance between the correlation peaks. If the synchronization output disappears, the data modulator will be based on previous sampling points. However, when the synchronization pulses are lacking, the data modulator is adjusted to the latest sampling estimate.

1.5 Processing gain of spread spectrum communication system

The signal/noise ratio at the output of the matched filter at the sampling instant $t = pT_c$ takes the following form for stationary noise:

$$(S/N)_{out} = \frac{E^2}{\int_{-\infty}^{\infty} N(f)|H(f)|^2 \, df} \tag{1.16}$$

where

$$E = \int_0^{T_d} [Aa(t)]^2 \, dt = \int_{-\infty}^{\infty} |H(f)|^2 \, dt$$

$a(t)$ is an arbitrary spreading code of ± 1 and A includes the path attenuation, assumed constant over the transmission interval. $H(f)$ is the Fourier transform of the filter impulse response $h(t)$, and $N(f)$ is the noise spectral density. If the noise is white with spectral density $N_0/2$ for all frequencies, eqn 1.16 reduces to

$$(S/N)_{out} = 2E/N_0 \tag{1.17}$$

The equation shows that the matched filter output signal/noise ratio, for a certain white noise level at the input, only depends on the data bit energy at the filter input. The exact shape of the signal to which the filter is matched does not influence the result. If the noise bandwidth of the filter is called $W = 1/T_c$, then the effective signal/noise ratio at the input is

$$(S/N)_{in} = \frac{P_{in}}{N_{in}} = \frac{A^2}{N_{in}} = \frac{E(1/T_d)}{N_0 W} \tag{1.18}$$

Using this in eqn 1.17 gives

$$(S/N)_{out} = 2T_d W (S/N)_{in} \tag{1.19}$$

At first glance this equation seems to indicate that the output signal/noise ratio depends on the time–bandwidth product, contradicting eqn 1.17. The explanation is that if the signal energy is kept constant the input signal/noise ratio will be inversely proportional to $T_d W$.

However, to obtain rejection against narrowband interference, to conceal transmission and to resolve multipath, a large time–bandwidth product is very important. $T_d W$ is therefore usually called the processing gain (*PG*) of the spread spectrum system.

If the interference appears in the form of a continuous wave

$$I(t) = I\cos[2\pi f_1(t)], \quad \text{then}$$

$$I(f) = \frac{I^2}{4}\delta(f - f_1) + \frac{I^2}{4}\delta(f + f_1) \tag{1.20}$$

and

$$(S/I)_{out} = \frac{2E^2}{I^2|H(f_1)|^2} \tag{1.21}$$

$|H(f)|^2$ is the power spectrum for the spreading code to which the filter is matched.

Using the Wiener-Kintchin theorem it is possible to obtain the power spectrum of the spreading codes:

$$|H(f)|^2 = \int_{-\infty}^{\infty} R(\tau)\exp(-j2\pi f\tau)\,d\tau \tag{1.22}$$

where $R(\tau)$ is the autocorrelation of the spreading code. The periodic autocorrelation function for m-sequences which are often used as spreading codes, can be graphically resolved as shown in Fig. 1.8. Then

$$R(\tau) = \sum_{n=-\infty}^{\infty} \alpha(\tau + npT_c) + \beta(\tau) \tag{1.23}$$

where n is an integer number of bits displaced. Inserting eqn 1.23 in eqn 1.22 gives

$$|H(f)|^2 = \left[\sum_{n=-\infty}^{\infty} \exp(jnpT_c 2\pi f)\right] \times \int_{-\infty}^{\infty} \alpha(\tau)\exp(-j2\kappa f\tau)\,d\tau - \frac{1}{p}\delta(f) \tag{1.24}$$

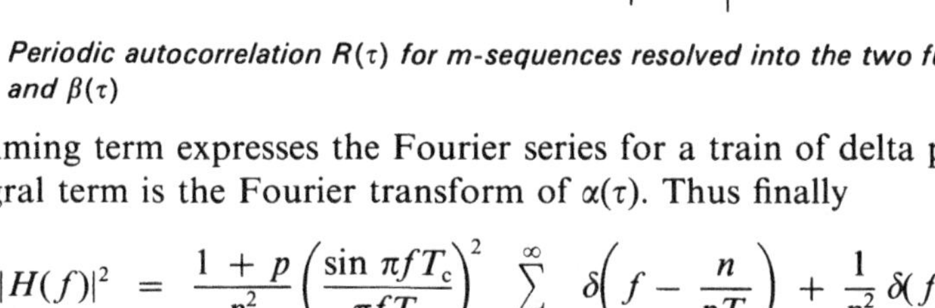

Fig. 1.8 *Periodic autocorrelation $R(\tau)$ for m-sequences resolved into the two functions $\alpha(\tau)$ and $\beta(\tau)$*

The summing term expresses the Fourier series for a train of delta pulses and the integral term is the Fourier transform of $\alpha(\tau)$. Thus finally

$$|H(f)|^2 = \frac{1+p}{p^2}\left(\frac{\sin \pi f T_c}{\pi f T_c}\right)^2 \sum_{\substack{n=-\infty \\ n\neq 0}}^{\infty} \delta\left(f - \frac{n}{pT_c}\right) + \frac{1}{p^2}\delta(f) \tag{1.25}$$

The envelope of this function is a $\sin^2 x/x^2$ function and the first zero crossing occurs for $f = 1/T_c$.

If a crude approximation is made and $|H(f)|^2$ is said to be constant across the noise bandwidth $W = 1/T_c$ and zero outside, then

$$|H(f)|^2 = \frac{P}{W} = \frac{E}{T_d W} \tag{1.26}$$

Inserting this in eqn 1.21 gives

$$(S/I)_{out} = \frac{2E}{I^2} T_d W \tag{1.27}$$

This equation shows that if the signal energy transmitted and the interference power are kept constant, the output signal/noise ratio will be proportional to the time-bandwidth product. Eqn 1.27 is frequently used to show how the effect of the interfering signal is reduced by the processing gain.

For active correlation systems this is easily visualized. The interference is spread by the local code generator at the receiver to a bandwidth defined by the code rate, and then the final filter matched to the data rate will reject all of the interference lying outside the filter band.

No such spreading takes place in a matched filter. In fact the power spectrum of the *m*-sequence just calculated is only valid for a periodic sequence while the matched filter is matched to one period or part of a period of the spreading code. The response of the matched filter to a continuous wave at the input will, since the filter is linear (neglecting possible quantization effects), be a continuous wave at the output, but with a different amplitude and phase. The amplitude will strongly depend on the interference frequency, but also on the code and its 'phase' as it appears in the filter realization.

The expected average filter output for a processing gain of p can be found using Parseval's theorem connecting signal energy in the time and frequency domain:

$$\sum^{p} a(j)^2 = \frac{1}{2\pi} \int_0^{2\pi} |A(\omega)|^2 \, d\omega \tag{1.28}$$

where $a(j)$ is the binary spreading code taking the values ± 1. Thus

$$2\pi p = \int_0^{2\pi} |A(\omega)|^2 \, d\omega \tag{1.29}$$

If $|A(\omega)|$ has an average amplitude k, then

$$\begin{aligned} 2\pi p &= 2\pi k^2 \\ k &= \sqrt{p} \end{aligned} \tag{1.30}$$

This is the amplitude value that a continuous wave excitation should give at the output of the matched filter if the $(S/I)_{out}$ is to be improved by the processing gain. However, as shown in Fig. 1.9, for a full period of the spreading code in each data bit large variations around $\sqrt{p}$ will occur for varying frequency, code and its phase. In fact, an averaging of the filter response, consecutively matched

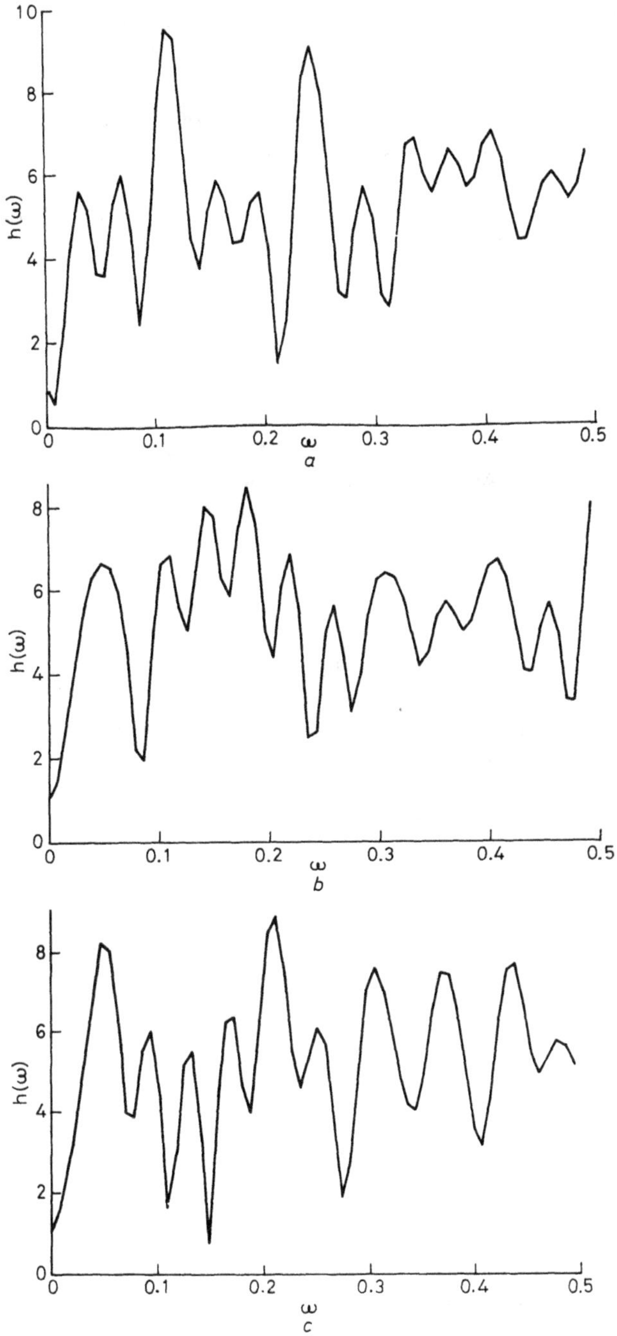

Fig. 1.9 *Impulse response for matched filters for codes with period 31. The impulse response is shown for the same m-sequence, p = 31, but for three different phase shifts*

to all possible phases of an *m*-sequence, gives a value $\sqrt{p}$ across the band (Fig. 1.10). The results show the danger of expecting any narrowband interference to be reduced by what is usually defined as the processing gain of the system. For *m*-sequences and a full period of the spreading code in each data bit the response at DC is fixed and equal to 1. For *m*-sequences and the full period spanning more than one data bit the response at DC will vary from data bit to data bit owing to changes in the ± 1 imbalance. In addition the rest of the

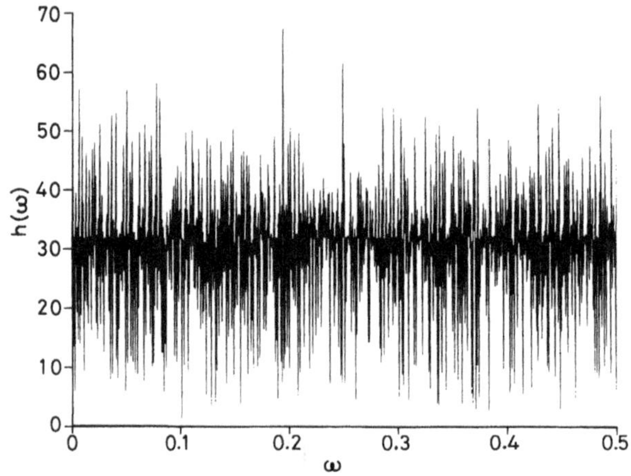

Fig. 1.10 *Frequency response of a matched filter for a m-sequence of period 1023 with generating function* $f(x) = x^{10} + x^3 + 1$

response will also change from data bit to data bit, causing an averaging effect at any particular frequency. A similar effect is of course present if the filter is matched to a full period but the code is changed from data bit to data bit. The last two situations do however require a matched filter which can be continuously programmed during data reception. The effect of tone interfering signals on system performance is analysed in detail by Milstein *et al.* [2] and Lunayach [3].

References

1 TURIN, G. L.: 'An introduction to digital matched filters', *Proc. IEEE*, 1976, **64**, (7), pp. 1092–1112.

2 MILSTEIN, L. B., DAVIDOVICI, S., and SCHILLING, D. L.: 'The effect of multiple-tone interfering signals on a direct sequence spread spectrum communication system', *IEEE Trans.*, 1982, **COM-30**, (3), pp. 436–446.

3 LUNAYACH, R.: 'Performance of a direct sequence spread-spectrum system with long period and short period code sequences, *IEEE Trans.*, 1983, **COM-31**, (3), pp. 412–419.

Chapter 2

Spreading techniques – unified description

2.1 Descriptors of carriers of information

The aim of this chapter is to provide an understanding of the mathematics and formalism required to describe signals at various stages in a communication link, and to introduce the degrees of freedom which are available to the intelligent user of electromagnetic waves. Waveform coding, modulation, frequency spreading and so on should be regarded as subsets of the spectrum of techniques that can be utilized for transmitting messages securely through space and even hostile environments. All these techniques deal with processing and coding of a *real time series*. This should be kept in mind when this time series, after the application of some mathematical tool, appears as a set of frequency spectra, narrowband complex representations or correlation functions.

In a communication link, these operations are most often performed when a signal passes through the various modulation stages in a transmitter, is mapped into one or a set of variables of the electromagnetic wave, and is carried in various directions in space until a print of this wave is pushed into a receiver. To fully describe this process the complete representations of the electromagnetic wave have to be introduced, namely the polarization properties and the spatial distribution of transmitting and receiving elements.

Fig. 2.1 gives a schematic picture of the various stages in the communication system. The source information I_1 is imposed on the electromagnetic carrier through a modulation process and then transferred to open space by means of a set of radiating elements. This radiation process introduces a set of new domains in addition to the time domain. The theoretical aspects of this phenomenon will not be covered in detail here, but some of the possibilities will be presented. Any more complete analysis of communication networks should include coding in the new domains. At present the space domains are only utilized to the extent that is obvious from a technical point of view, but these domains are evidently available for coding in very much the same way as is the time domain.

Any transmitter or receiver will give certain characteristics to the polarization and the angular and spatial distribution of the electromagnetic field. These characteristics will have to be included as part of the total description of the system. Several very interesting features can be achieved by allowing these characteristics to be time variables.

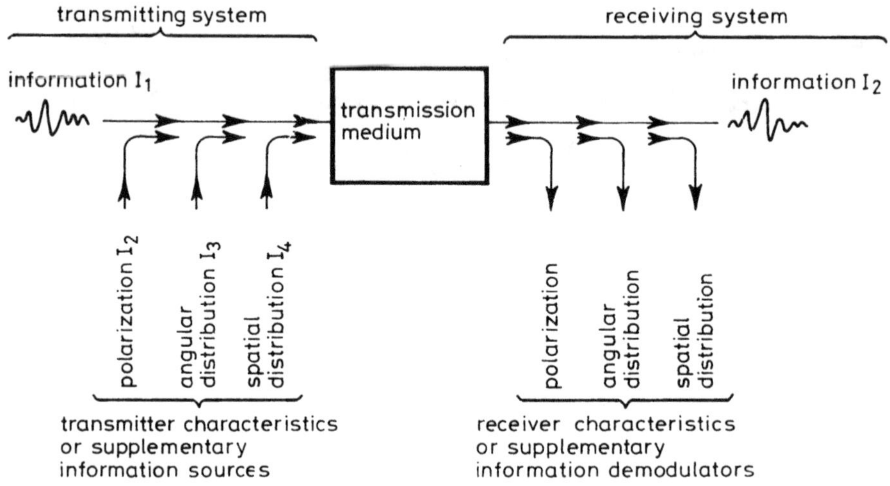

Fig. 2.1 *Physical properties of the electromagnetic wave introduces added information to the transmitted message*

The polarization state of the electromagnetic wave takes advantage of the transversal nature of the electromagnetic wave, by which two independent domains are immediately introduced. It is well known that an antenna emits a certain polarization of the electromagnetic wave, but usually the two transversal dimensions are not exploited independently. Interference rejection or communications security can be achieved by manipulating the polarization properties.

Similarly it can be argued that when the emissions from a set of radiating elements are superimposed to give an angular distribution of radiated field strength, it is possible to consider a system with individually coded radiation elements. The traditional direction antenna could be replaced by a set of independent beams, and individual coding would give an improvement when a network of receivers was available.

On a larger scale, a network of stations permits spatial processing and focusing. This degree of freedom is of use when it is required to solve a specific cell in space. This is the inverse of the synthesis of the coded antenna pattern: the electromagnetic field in space is probed by a set of units to deduce the angular distribution of radiation. For example, a jammer can be eliminated or a concealed source detected using spatial focusing.

Owing to the reciprocity of the electromagnetic waves, these operations can be performed equally well at the receiving station or stations.

2.2 Electronic warfare scenario

Consider a scenario involving a friend and a foe. The friend establishes his communication network, and it is his duty to keep this network operational. He has to face electronic warfare: the enemy will ultimately establish a set of measures to deny the friend his objectives, or will tap his communication lines and convey this information to higher levels in his hierarchy.

In this dynamic situation, and assuming that both parties do their best, we can define the objectives of the owner of the communication network, the friend, as

1 To establish and maintain a communication network.
2 To counteract any attempt by the enemy to inhibit or exploit (detect, eavesdrop on) the communication flow.

The enemy has the following challenges:

1 To detect and/or localize the existence of communication links. This is always the first step: steps 2–5 may follow.
2 To eavesdrop on the information flow.
3 To block the information flow (jamming).
4 To insert false information (spoofing).
5 To choose new strategies following any countermeasures by the network owner.

When the two systems interact a dynamic situation will arise: the communication or electronic measures (EM) will be subjected to electronic countermeasures (ECM), which will in turn trigger electronic counter countermeasures, and so on. The situation is illustrated as follows:

Actions by friend	*Actions by foe*
EM	
	ECM
ECCM	
	$EC^{n-1}M$
$EC^{n}M$	
	$EC^{n+1}M$

The limiting factors in this process are time and cost. Included in these parameters are such elements as technological improvements and conceptual advancements. These will control the growth of the process.

To give a more complete picture, the enemy's communication network should also be considered. Assuming the parties use the same strategy, the following situation arises:

Actions by friend	*Actions by foe*
EM	EM
ECM	ECM
ECCM	ECCM
$EC^{m-1}M$	$EC^{n-1}M$
$EC^{m}M$	$EC^{n}M$
$EC^{m+1}M$	$EC^{n+1}M$

At levels n and m, each player will have to distribute his resources between actions taken to maintain his own network and actions taken to deceive the foe's network. The effort expanded on these tasks will be evaluated by superiors, and as a consequence the processes can be halted at levels different for the two parties. If the friend has been instructed to concentrate his efforts upon the survival of his own network, and the enemy is spending his resources on actions threatening the friend's network rather than on securing his own, then the friend will at a certain time be furnished with a more sophisticated network than the enemy.

To analyse this situation, the concept of cost effectiveness of countermeasures has to be introduced. Assume that a player's network at some time has condition EC^nM, where n is an even number. This implies that the network has been sophisticated in $n/2$ steps, to fight another $n/2$ steps of countermeasures from the enemy. To analyse this model, it is assumed that a unit step in the evolution of the process can be defined, implying that a practical system can jump any number of steps at any one instant.

When the most recent process is an action by the friend, it has not been challenged by the other player and the friend's network is consequently operational. On the other hand, when the most recent move has been made by the enemy, the network is nonoperational.

network operational $\Rightarrow$ EC^nM process in n even state

network not operational $\Rightarrow$ EC^nM process in n odd state

If R_n is the cost to the enemy of developing and deploying countermeasures to invalidate the friend's network in state EC^nM, and P_n is the cost to the friend of improving his system beyond the present capabilities of the enemy, the following situation arises:

Friend's activities	*Enemy's activities*
EC^nM	R_n
P_n	$EC^{n+1}M$

$EC^{n+2}M$

R_N	EC^NM
$EC^{N+1}M$	P_N
	$EC^{N+2}M$

This gives the following set of cost factors:

P_n	cost of friend to improve own net
P_N	cost of foe to improve own net
R_N	cost of friend to beat foe's net
R_n	cost of foe to beat friend's net

Assume now that the strategy of each player can be described by the proportion of expenditure used on defensive and offensive activities:

$$\eta_{\text{friend}} = P_n/(R_N + P_n)$$

$$\eta_{\text{foe}} = P_N/(R_n + P_N)$$

An index of $\eta = 1$ indicates an entirely offensive strategy with no emphasis on countermeasures, whereas η close to zero implies no strengthening of one's own net.

From this model, it is intuitively clear that the nets on either side can be sophisticated to an arbitrary level, dependent upon individual strategies and total effort.

Associated with a specific effort P or R is its effect. The effect is dependent upon the effort through the efficiency. Fig. 2.2 shows the cumulative effect of all

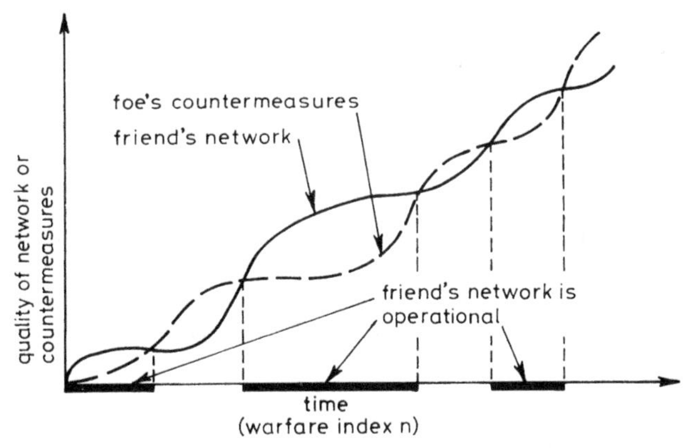

Fig. 2.2 *Dynamic game of electronic warfare*

efforts plotted against the steps in the war game or against time. This is done for measures as well as for countermeasures. When the quality of the countermeasures is greater than that of the owner's measures, the net is not useful. The cost efficiency of each step in this process will determine the delay (dwell time) before the enemy has developed sufficient countermeasures.

To form a complete picture, the variety of electronic measures on either side should be considered. Old communication systems will generally be operative, because a small capital cost will keep them cost effective even with a vanishing usefulness. Also, communication systems will not always be coordinated within a tactical group: this can be accredited to suboptimum control of own strategy or unforeseen events.

The result is that the degree of interference between communication links will be a very important factor. The situation can be illustrated as follows. A set of n friendly nets will have to operate in parallel, and each will be in a particular state in the warfare process. Net j will also have a certain effect $O_{j,i}$ on net i. Very often, assessments have to be made of the influence of a change in one net on the performance of the others. For example:

Net 1		*Net 2*
EM		EM
ECM		ECM
EC^mM	O	EC^nM
$EC^{m+1}M$	O'	$EC^{n+1}M$

In this situation, the problem is to determine the coefficients O and O'. O is the influence on net 1 in state m of net 2 in state n. O' is the corresponding influence if net 1 changes to state $m + 1$. A very simple illustration of this is the situation where the change in net 2 was an increase in transmitted power to overcome a traffic squeeze or jamming attempt from the enemy. This move obviously has an adverse effect on net 1. Hence, the O factor is also referred to as the *self-jamming factor*.

In the following, the technical and mathematical aspects of cost efficiency will be discussed. The efficiency of various scientific and technical advances will be considered from an information theoretical point of view. It should, however, be borne in mind that there are other cost factors in the war game – those governed by the overall system and not influenced by developments in the electronic warfare scenario. These factors, for example the costs of technology, administration and development, will not be discussed.

2.3 Electronic warfare techniques

Chapter 2.1 introduced the various domains available to the player of the electronic warfare game. A distinction was made between time and space domain operations, and as these operations can be performed in series the total benefit will be a function of all operations performed.

Specifically, given the efficiencies E_{time} and E_{space} for time and space domain operations, the total benefit will be the product of the two terms:

$$E_{system} = E_{time} E_{space}$$

Also, each of these elements will generally be the product of a set of independent operations in each domain:

$$E_{system} = \prod_n E_{time,n} \prod_m E_{space,m}$$

The efficiency E will be the efficiency of one specific operation. The operations discussed here are detection, location, eavesdropping and jamming, each by electronic means:

Detection to establish the existence of an information transfer.
Eavesdropping to tap the information flow within a net.
Jamming to block the information flow in a net.
Localization to establish the position of the information source.

Each of these operations refers to some specific situation, that is a position on the electronic warfare ladder. Whereas all four operations apply for the friend–foe relationship, only jamming (self-jamming) will be important for the interference of own nets.

Depending upon which of the system parameters one considers, E_{system} can be any one of the following:

$E_{detection}$
$E_{eavesdropping}$
E_{jammer}
$E_{localization}$

Then any one of these is the product of the efficiency factor of a series of operations. For example, the use of a jamming waveform with an efficiency E_{code} will give a system jamming efficiency of

E_{jammer}	$=$	E_{code}	match of jammer signal
		$\times\ E_{TX\,antenna}$	pattern of transmitting antenna
		$\times\ E_{RX\,antenna}$	pattern of receiving antenna

The first factor is a time domain operation and the second and third are space domain operations.

For military electronic warfare there is another aspect to be introduced. This is the amount of information about the opponent's communication system that is available to each player. This challenge is not peculiar to the electronic part of warefare; however, espionage and intelligence make electronic warfare particularly sensitive in this respect.

Each player will aim at keeping the opponent uninformed of his activities, systems and strategies. For electronic countermeasures, the intelligence data available can for analytical purposes be at one of three levels:

1 No information available.
2 The main system parameters are known.
3 In addition, the strategy and codes are known.

Level 3 implies that the enemy has full knowledge and is in the same position as the legitimate user of the net.

Depending upon the level of knowledge, the cost of a specific operation will vary. It is, for example, extremely important for a narrowband jammer to know the frequency as well as the position of a narrowband net. Furthermore, as will be shown later, if the information is protected by a code, the transmission cannot be acted upon under certain circumstances unless this code is known.

2.4 Time domain coding

This chapter introduces the operations that can be applied for coding in the time domain. Only one-dimensional coding will be discussed. This means that the processing will take a single time series through a set of operations and deliver a new time series to the next stage in the processing system. The basic idea of time domain coding is to add a certain amount of information (a code) to an input data stream, and in such a manner that the original data stream can only be decoded by knowing the original code. As an example, this can simply be accomplished by replacing the data bit by a code sequence. The ratio betwen code flow rate and data flow rate, which in Chapter 1 was referred to as the processing gain, should be as high as possible to give the best performance.

The use of an unknown spreading code gives the communication a certain privacy. In order for a third party to spoof the communication, not only does he need a replica of the spreading code used but also he must align his code in time with that of the legal transmitter. The reason for this is simply that the receiver will, as soon as it is synchronized to the legal transmitter, insert a time window and then only accept correlation peaks within this window. The intelligent jammer can however accomplish this task by 'sliding' his timing to capture the victim receiver. In a burst communication system, where communication links are established and disconnected on a random and irregular basis, there would be plenty of opportunities for an unauthorized user with knowledge of the spreading codes to spoof the system. Thus certain procedures for constructing a legal data message may be needed. Based on these procedures, legal and illegal messages could be sorted out at the data handling level. However, in many military applications it is also considered necessary to design the code generator so as to prevent an unauthorized listener deriving the code design. This can be achieved by special nonlinear code generators or by changing the spreading codes, taken from the families of codes described in Chapter 3, on a random and irregular basis during a message transfer. Thus any illegal transmitter would have difficulties in spoofing the system for any significant period. In a commercial system, without the ECM effort, changing the spreading codes from time to time would be considered to give the privacy necessary for most occasions.

The process of imprinting codes on a data stream in a reversible manner requires that the data flow time series is mapped into a multidimensional space with a sufficient number of free variables. Any suitable transform could be applied, but as the communication and radio engineer is assumed to be attached to the theory of frequency analysis, this will be chosen as the basis for the treatment. As an introduction and recapitulation, an outline will be given of how time-frequency representation of signals describes a time series in a manner that is fairly simple yet gives all the degrees of freedom needed.

2.4.1 *Time–frequency maps*

The Fourier transform of a time series $a(t)$ is given as

$$A(f) = \int_{-\infty}^{\infty} a(t) \exp(-j2\pi ft)\, dt$$

where f is frequency. We also have

$$a(t) = \int_{-\infty}^{\infty} A(f) \exp(j2\pi ft)\, df$$

$A(f)$ and $a(t)$ are referred to as the *Fourier pair.*

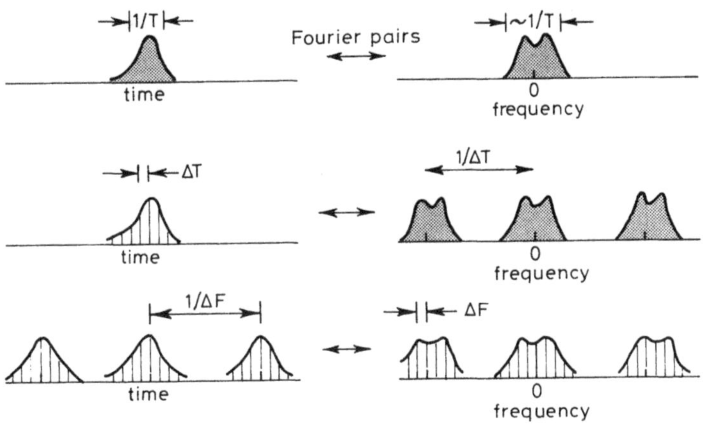

Fig. 2.3 *Sampling in the time domain corresponds to replication in the frequency domain and vice versa*

Furthermore, sampling in one domain corresponds to a replication in the other domain with an interval equal to the inverse of the sampling interval. This is reciprocal in the two domains, and a sampled and replicated function in one domain implies a replicated and sampled function in the other domain (Fig. 2.3).

From Fourier theory it is known that a multiplication in one domain equals a convolution in the other. This implies that the spectrum of a windowed sinusoidal signal is a sin x/x function. This applies for a square window, which gives the best spectral resolution at the cost of high sidelobes.

There are two important aspects of windowing. The first is the influence on spectral resolution and sidelobes, and the second is how the data set for the Fourier analysis is chosen along the time axis.

The input signal $a(t)$, an infinite time series, is partitioned into a set of finite time series $a_i(t)$ taken at intervals Δt with the window $w(t)$:

$$a_i(t) = a(t)w(t)$$

The total sum of all subseries

$$a'(t) = \sum_i a_i(t) = \sum_i a(t)w(t)$$

will only be equal to the original time series in the case where

$$\sum_i w(t - i\Delta t) = 1$$

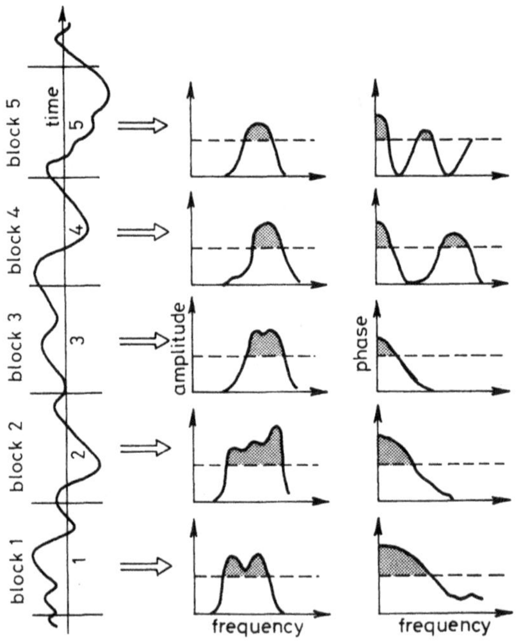

Fig. 2.4 *Illustrations on how successive time slots are represented by complex spectra* (*dotted lines are chosen thresholds for one-bit quantization*)

This is obviously the case for a rectangular window with width Δt, as illustrated in Fig. 2.4. This gives a set of frequency spectra with a sin x/x response, leading to a spectral -3 dB or 50% resolution equal to $\Delta f = 1/\Delta t$.

It should be observed that overlapping windows give redundant information. Consider a window the width of which is m times the window separation of two successive windows:

$$w(t) \quad = \quad \text{rect}\ (x/m\Delta t)\ \frac{1}{m}$$

This window also satisfies the criterion

$$\sum_i w(t - i\Delta t) \quad = \quad 1$$

However, each value of the time series $a(t)$ will now be used m times, giving an m-fold redundancy of the frequency spectra.

The time–frequency representation corresponding to the amplitude and phase time variation of the signal in Fig. 2.4 is shown in Fig. 2.5. The frequency spectra are represented by one-bit quantization for amplitude and phase, and a two-dimensional diagram with time as the horizontal axis and frequency as the vertical axis is introduced. Other approaches give multilevel quantization or even three-dimensional pictures. Most applications in this book do not require more than one quantization level. One square in the map of Fig. 2.5 represents one bit of information if $\Delta f \Delta t = 1$.

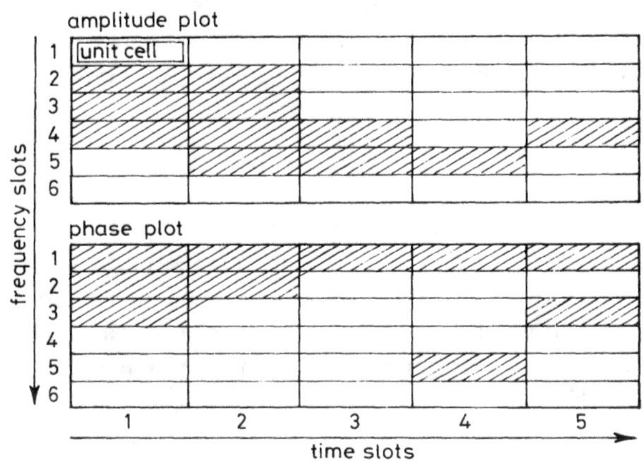

Fig. 2.5 *Amplitude and phase plot as derived from the time series in Fig. 2.4*

Windows of shape other than square can be useful where different requirements are set for the resolution in either domain. The rule is that the better the resolution in one domain, the poorer it will be in the other. As an example, a triangular window of width $2\Delta t$

$$\begin{aligned} w(t) &= \text{triangle}\ (t/\Delta t) \\ &= \text{rect}\ (t/\Delta t) * \text{rect}\ (t/\Delta t) \end{aligned}$$

gives a spectrum response characterized by the frequency transform $W(f)$ of the window function $w(t)$:

$$W(f) = \text{FT}[w(t)] = \text{sinc}^2(f\Delta t)$$

It should be observed that this triangular window also obeys the criterion

$$\sum_i w(t - i\Delta t) = 1$$

Unequal emphasis is put upon each value of $a(t)$ within each spectrum, but the total emphasis averaged over neighbouring spectra is constant.

A final remark should be made on the issue of alternative spectral estimators. It is obvious that given *a priori* knowledge of the spectrum, an improvement in the spectral resolution can be obtained. As an example, the maximum entropy spectral analysis method of order n calculates the position and amplitude of a total of n sinusoids with a resolution only given by the accuracy of the calculations. This is valid as long as the input signal obeys the assumptions of the model. However, if the number of sinusoids exceeds that of the model, the process does not converge. This might be the case with a noisy signal, where the noise represents a very large number of sinusoids.

In the following, time–frequency maps are described by the time–frequency function $C(t, f)$. The variables t and f here represent discrete independent variables, and the total spans of these variables are, for a certain code, T and F respectively. This implies that the total number of elements in the map is TF. Hence the processing gain of a code as represented by a time–frequency map is simply:

$$\text{processing gain} = TF$$

2.4.2 *Principles of various spreading techniques*

On the basis of time–frequency maps, various methods for coding will be presented and discussed. Fig. 2.6 illustrates the basic idea of coding, namely to substitute a data bit with a code matrix. The figure shows a general code pattern in which each of the time–frequency slots is independent.

As illustrated in Fig. 2.7, the ratio T/F can be chosen to suit different needs. The frequency width F can be chosen to suit a band allocation scheme, or the duration T can be chosen to give a required data rate.

In the following, a set of coding methods will be introduced. The methods are all represented by their frequency–time maps. The choice of modulation of each frequency–time slot is not discussed here, as that would lead to more detailed analysis and is not essential to understand the basic idea of the coding schemes. The modulation can be a simple on/off amplitude modulation or phase shift modulation or more complex multilevel modulation.

2.4.2.1 *Generalized coding scheme*

This implies that no restraints are imposed on the structure of the code, and the C map will have the random structure as seen in Fig. 2.8. From this it follows that there is full freedom to fight countermeasures or transmission medium limitations (e.g. multipath).

In this case, a time–frequency map descriptor (amplitude or phase or both) is modulated by a two-dimensional code:

$$C(t, f) \quad = \quad a(tf)$$

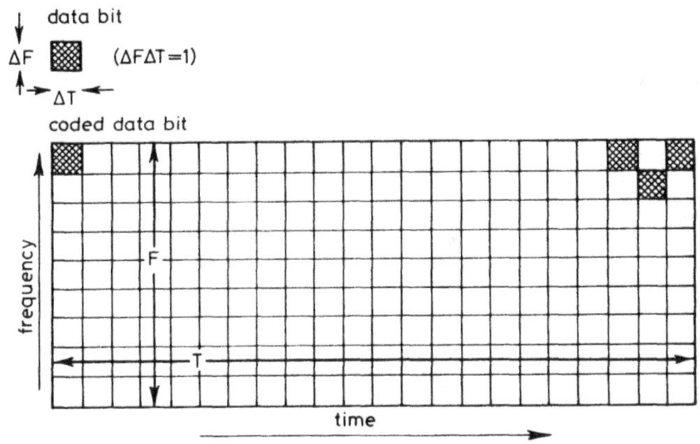

Fig. 2.6 *By coding, each data bit carrying one unit of information is represented by a coded waveform carrying TF units of information*

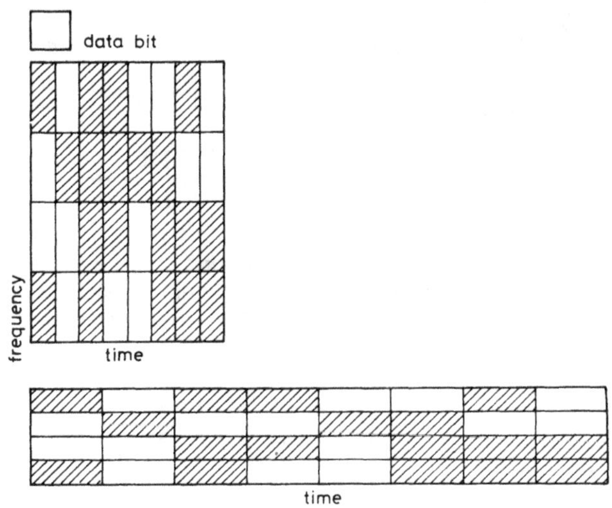

Fig. 2.7 *Relative T/F ratio of any code can be adjusted to match bandwidth capability of channel or data transmission rate requirements*

2.4.2.2 *One-frequency hopper*

In this scheme only one frequency slot is activated at a time, and the code selects which slot is to be activated (Fig. 2.9):

$$C(t, f) = \begin{cases} 1 \text{ for } f = a(t) \\ 0 \text{ otherwise} \end{cases}$$

when $a(t)$ is the code function and span $[a(\ )]$ is equal to the span of f.

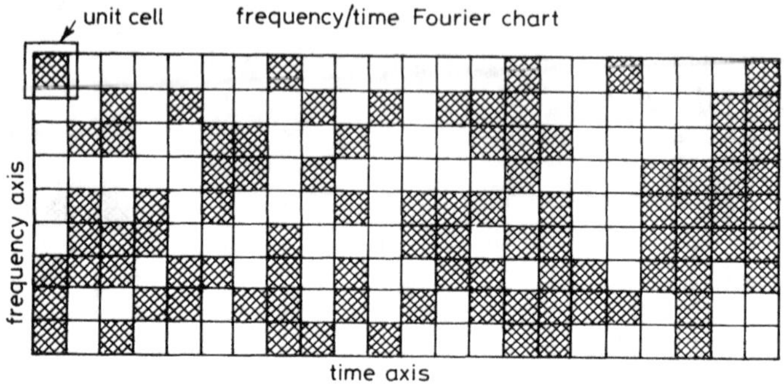

Fig. 2.8 *Generalized coding scheme with independent coding along the frequency and time axis. The example shows a code with TF = 9 × 22 = 198 elements*

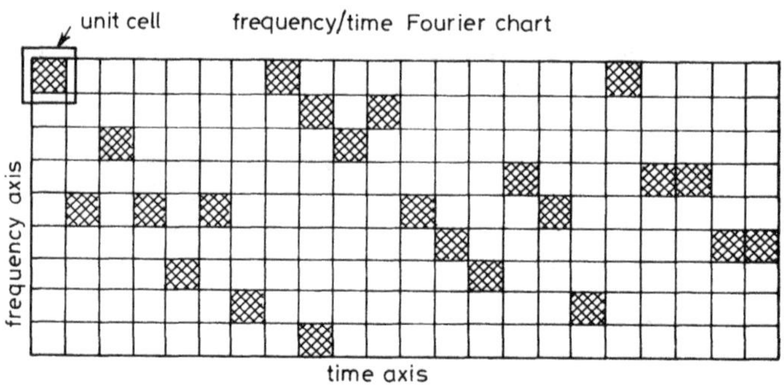

Fig. 2.9 *One-frequency hopper coding scheme. Note that compared with the generalized scheme in Fig. 2.8, the processing gain is reduced by a factor 9 corresponding to the number of noncoded elements along the frequency axis PG = 22*

2.4.2.3 *Multifrequency hopper*

Here the C function is allowed to be activated in a number of frequencies at one time, governed by a total of i codes $a_i(t)$ ($i \ll F$) (Fig. 2.10):

$$C(t, f) = \begin{cases} 1 \text{ for } f = a_1(t) \text{ or} \\ \quad f = a_2(t) \text{ or} \\ \quad \cdot \\ \quad \cdot \\ \quad \cdot \\ \quad f = a_i(t) \\ 0 \quad \text{otherwise} \end{cases}$$

This is an improved version of the single-frequency hopper, and the resulting processing gain is improved.

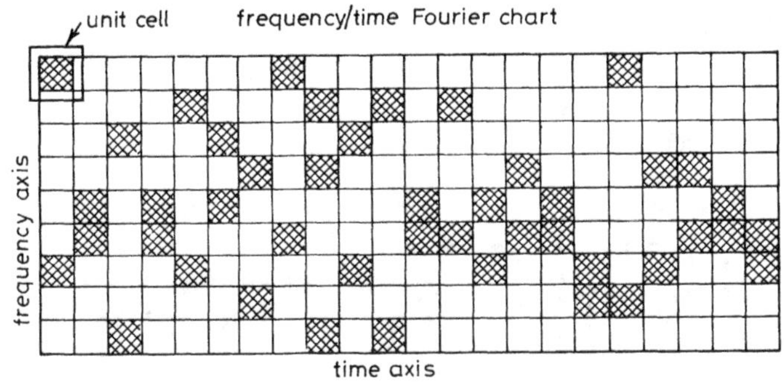

Fig. 2.10 *Multifrequency hopper, here represented by a two-frequency hopper. In this case the processing gain has improved from 22 as in the simple hopper case in Fig. 2.9 to 2 × 22 = 44*

2.4.2.4 *Time hopper*

If only the time axis is considered, one can allow an uncertainty in the time of arrival rather than an uncertainty in the frequency (Fig. 2.11). Given a time interval T, the code function will be

$$C(t, f) = \begin{cases} 1 \text{ for } t = a \\ 0 \text{ otherwise} \end{cases}$$

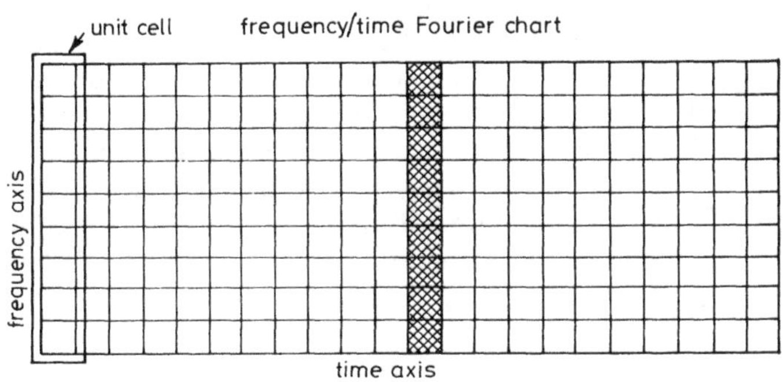

Fig. 2.11 *Time hopper approach. Note that the time–frequency unit cell is chosen rectangular rather than quadratic to conform with the standard grid. This system has PG = 1*

In such a system, it would be advisable to let the C function represent phase rather than amplitude to increase the total power transmitted.

2.4.2.5 *Direct sequence system* (multitime hopper)

To increase the information content of the code, a set of time slots can be set in the C function (Fig. 2.12). In the case where every slot is manipulated, the direct sequence system is obtained. Here, the state of each slot is given by the code

$$C(t, f) = a(t)$$

It should be noted that in Figs 2.11 and 2.12 the system unit cell is chosen to be rectangular rather than square to fit the code in a C function with the same T/F ratio as the frequency domain codes.

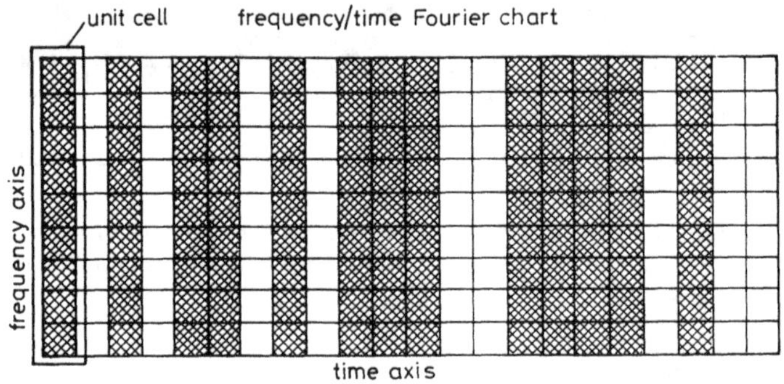

Fig. 2.12 *Multitime hopper or direct sequence coding scheme. This system gives full coding along the time axis (PG = 22)*, but no independent structuring along the frequency axis

The conceptual relationship between direct sequence, frequency hopping and hybrid schemes is shown in Fig. 2.13 for a comparable data rate and transmission security. The direct system spreads the transmitted energy over the full frequency band by using a rapid code, and transmits one data bit at a time. The frequency hopper transmits a number of data bits while dwelling on one frequency, and then jumps to another frequency slot and continues transmission. In the hybrid system, the signal energy in each frequency slot is protected by a direct sequence code, thereby broadening the width of the frequency band activated at any time. In Fig. 2.13 the various coding schemes are given the same total processing gain, so that the conceptual characteristics of the coding methods can be easily visualized.

2.4.3 *Countermeasures assuming no system knowledge*

Whatever the type of countermeasures – jamming, detection or eavesdropping – the efficiency of the operation will depend upon the match of the time–frequency map of the 'attack' waveform to the map of the signal. An assumption will have to be made concerning the degree of knowledge of the system being attacked, and (using the distinctions made in Section 2.3) the states of full knowledge, system knowledge and no knowledge will be analysed.

In addition, it is of vital importance that the time reference is known. It is evident that synchronization is important, as the matched filter is matched only within one time slot.

If the interrogating waveform's time–frequency map is $I(t, f)$, the efficiency or 'match' between this and the code function is defined as the *success factor* X_{CI}. In the general case, the I function will not have the same span in the time and

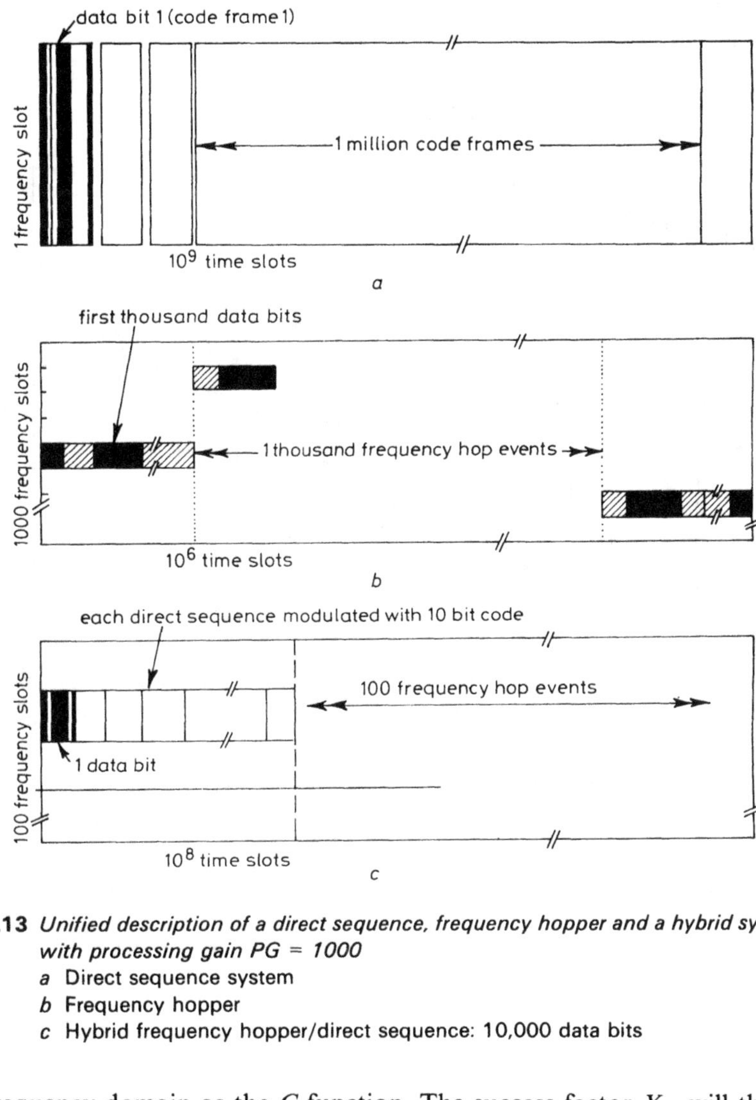

Fig. 2.13 *Unified description of a direct sequence, frequency hopper and a hybrid system, all with processing gain PG = 1000*
a Direct sequence system
b Frequency hopper
c Hybrid frequency hopper/direct sequence: 10,000 data bits

the frequency domain as the C function. The success factor X_{CI} will therefore have to be defined in the mean, averaging over every possible absolute position of the C and I functions:

$$X_{CI} = \sum_t \sum_f C(f, t) I(f, t)$$

$$C(f, t) * I(f, t)$$

There are three contributions to the data security (Fig. 2.14). The first is the processing gain of the code, which is equal to the number of independent elements in the data C function:

$$PG = FT$$

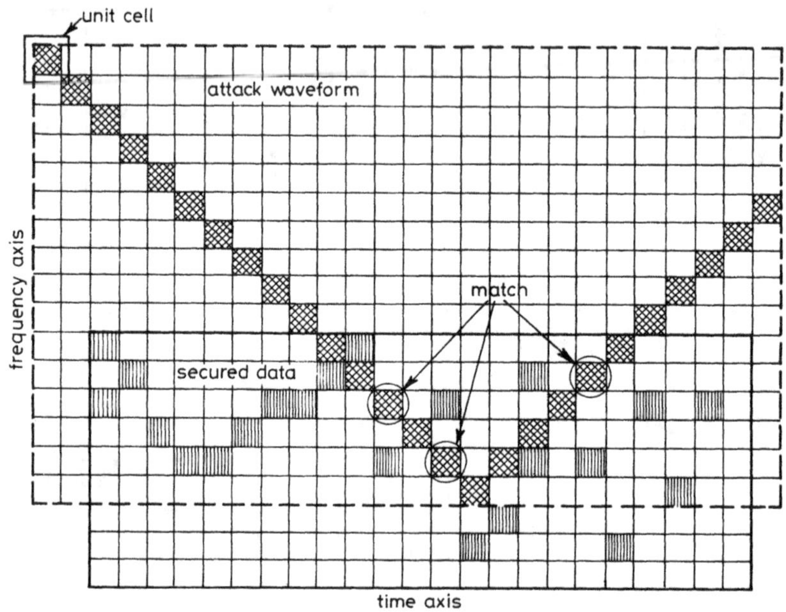

Fig. 2.14 *When there is no knowledge available to the designer of the attack waveform, he will be forced to distribute his resources within a vast area in the time–frequency diagram*

Secondly, there is the uncertainty of the time reference. Assuming that the uncertainty in synchronization is δT_{sync}, this gives an added security:

$$\frac{\delta T_{sync}}{\Delta T}$$

(ΔT and ΔF are the unit elements in the time–frequency diagram.)

Thirdly, the uncertainty along the frequency axis (δF) gives the frequency uncertainty improvement:

$$\frac{\delta F}{\Delta F}$$

This gives the total efficiency factor assuming full system knowledge:

$$E_{time} = FT \frac{\delta T_{sync}}{\Delta T} \frac{\delta F}{\Delta F}$$

This may be compared with the efficiency factor of $E_{time} = 1$ for a system given no system knowledge.

2.4.4 *Countermeasures assuming system knowledge*

This implies that the total span along the time and frequency axes is known, but that no information as to the structure of the waveform within this frame is available. This is a common level of knowledge. Frequency allocation and data formats are often not classified and are thus fully available.

The data security will here depend on the frequency uncertainty factor:

$$E_{\text{time}} = \frac{\delta F}{\Delta F}$$

This situation is illustrated in Fig. 2.15. Those elements in the time–frequency map are circled in which there is coincidence of the data and attack waveforms.

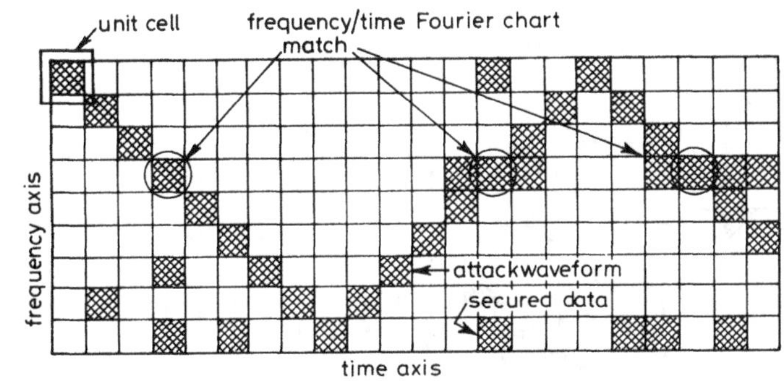

Fig. 2.15 *When the main system parameters are known, high possibility for some degree of match between codes can be obtained*

2.4.5 *Countermeasures assuming full system knowledge but no synchronization*

This situation may arise either if the legitimate receiver has failed in synchronization, or if an illegitimate receiver does not have the information required to obtain synchronization. Synchronization is most often achieved by transmitting a code with sufficient complexity to ensure secure detection even if the time reference is not known in advance of the synchronization process. The time reference can also be established by other means, such as by parallel communication links. High precision time references (atomic clocks) can also provide initial time information, which may be close to sufficient for synchronization, and thereby reduce the time uncertainty δT_{sync}.

High data rate systems require synchronization to better than the duration of one time slot (code chip ΔT). This might scale to distances of only a few centimetres (a 1 GHz chip rate gives a 30 cm required position accuracy). For mobile systems, synchronization must therefore be acquired at intervals within which the mobile unit has moved only a fraction of a 'chip wavelength', or position or velocity has been measured. By tracking one of these parameters, corrections to the time reference can be estimated.

The data security factor for the unsynchronized system will be equal to the synchronization uncertainty:

$$E_{\text{time}} = PG \frac{\delta F}{\Delta F}$$

The code processing gain is present because the code is now broken. This demonstrates that a spread spectrum matched filter has considerable immunity to countermeasures even if the code is accessible, because the synchronization offers additional security.

Fig. 2.16 illustrates this. The attack waveform is not aligned along the time axis, and only spurious coincidences of the attack code and data code occur.

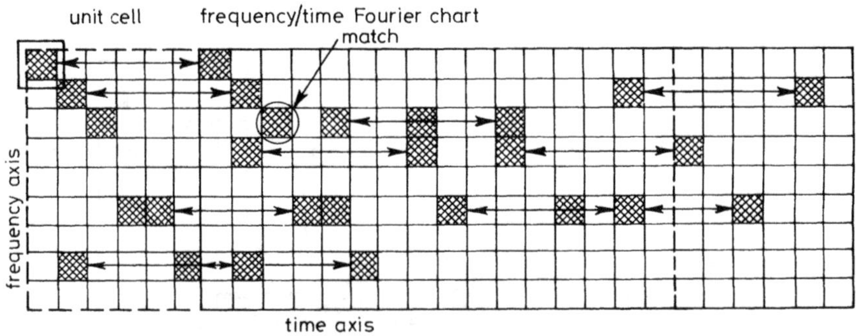

Fig. 2.16 *When the code is known, it is still essential to know the time axis reference points (synchronization)*

2.4.6 *Synchronized system*

When the system is synchronized, no uncertainty exists about the location of the code on the time and frequency axes. If the code is known as well, the time domain security has its maximum value:

$$E_{\text{time}} = PG \frac{\delta T_{\text{sync}}}{\Delta T} \frac{\delta F}{\Delta F}$$

This situation is represented in Fig. 2.17.

In summary, the data security factor for time domain processing for the various assumptions is presented in Table 2.1.

Consider a typical system with the following parameters:

$FT = 1000$		processing gain
$\frac{\delta T}{\Delta T} = 1000$		uncertainty in time equals data bit length
$\frac{\delta F}{\Delta F} = 10$		band allocation may be 1 out of 10

From these formulas it is seen that, with the net intact and functional, the enemy is a factor of 10^7 worse off in using the net illegally. This factor may, depending upon the approach chosen, lead to one of the following;

An increase in cost of 10^7
An increase in processing time of 10^7
An increase in processing complexity of 10^7; etc.

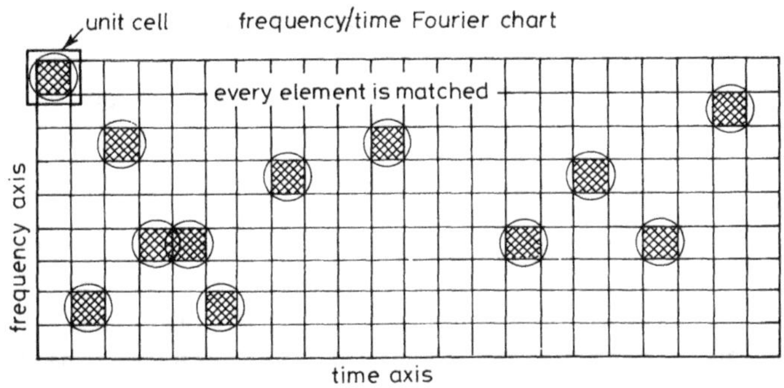

Fig. 2.17 *Perfect match is obtained in a synchronized system with open codes. This implies that the data is no longer secure*

Table 2.1

	E_{time} equals	Applicability to friend	Applicability to enemy
No system knowledge	1	None	The net is not accessible
Basic system knowledge	$\frac{\delta F}{\Delta F}$	None	Transmission frequencies are known
Synchronized system and basic system knowledge	$\frac{\delta F}{\Delta F}\frac{\delta T_{sync}}{\Delta T}$	Operation when code is not valid	Code is not known
Unsynchronized system with known code	$PG\frac{\delta F}{\Delta F}$	Synchronization phase	The synchronization process is not known
Synchronized system with known code	$PG\frac{\delta T_{sync}}{\Delta T}\frac{\delta F}{\Delta F}$	Normal operation	The net is broken

2.5 Space domain coding

The space domain gives in principle four independent axes for coding – the three volume space axes, plus the polarization properties of electromagnetic waves.

The most important aspects of space coding are those associated with beam forming antenna concepts and spatial focusing; the polarization offers another dimension which can be utilized for jammer avoidance etc.

2.5.1 *Volume space coding*

The synchronization of transmitters and receivers in space facilitates the use of space domain techniques for electronic countermeasures. Jamming and eavesdropping can be avoided by the proper use of space domain operators, and of course the positioning of transmitters relies entirely on space domain techniques.

2.5.1.1 *Directional antennas*

Directive antennas have provided the main solution to interference rejection problems.

In the category of directional antennas come all the various methods for generating beams of radio energy in one preferred direction. If the distribution of the electromagnetic field over the antenna is $E_a(\boldsymbol{r})$, the antenna far field pattern $E(\boldsymbol{k})$ is given as the Fourier transform of the aperture function integrated over the antenna:

$$E(\boldsymbol{k}) = \int E_A(\boldsymbol{r}) \exp(-j\boldsymbol{kr})\, d\boldsymbol{r}$$

$\boldsymbol{k}$ is the electromagnetic field wave number, the angle of which is the propagation direction of that wave number component.

This implies that an antenna of width Δr gives a lobe width of the order of

$$\theta \sim \frac{\lambda}{r}$$

The exact lobe width as well as the sidelobe radiation pattern depend on the fine structure of the aperture function. Recent developments in radio frequency processing components have made it possible to construct antennas with time variable pointing angles and radiation patterns.

Specifically, it is possible to design antennas with deep nulls in the radiation pattern for a number of directions. This is highly desirable when there are interference sources located at discrete locations in space (see Chapter 6). The number of radiation nulls is equal to the number of discrete radiation elements minus 1. The depth of the nulls will be dependent on the bandwidth of the signal transmitted or received by the antenna, in the manner familar from filter theory.

The total benefit from the antenna pattern will directly equal the value of the radiation pattern in the direction in question:

$$E_{\text{space}} = E(\boldsymbol{k})$$

for $\boldsymbol{k}$ equal to directions to transmitter/receiver.

Fig. 2.18 gives an example of how the directionality of the antennas can provide protection against eavesdropping and interference (jamming). The transmitter points a directional beam towards the omnidirectional receiver. Eavesdropper 1 (dipole antenna) does not succeed because he is outside the transmitter beam. Eavesdropper 2 is known to the transmitter, and a main beam null is established in his direction. Jammer 1 points a highly directional beam towards the receiver, and jammer 2 radiates energy from a different direction. In this case, the antenna is capable of generating one null only and cancels the stronger jammer; jammer 2 cannot be avoided.

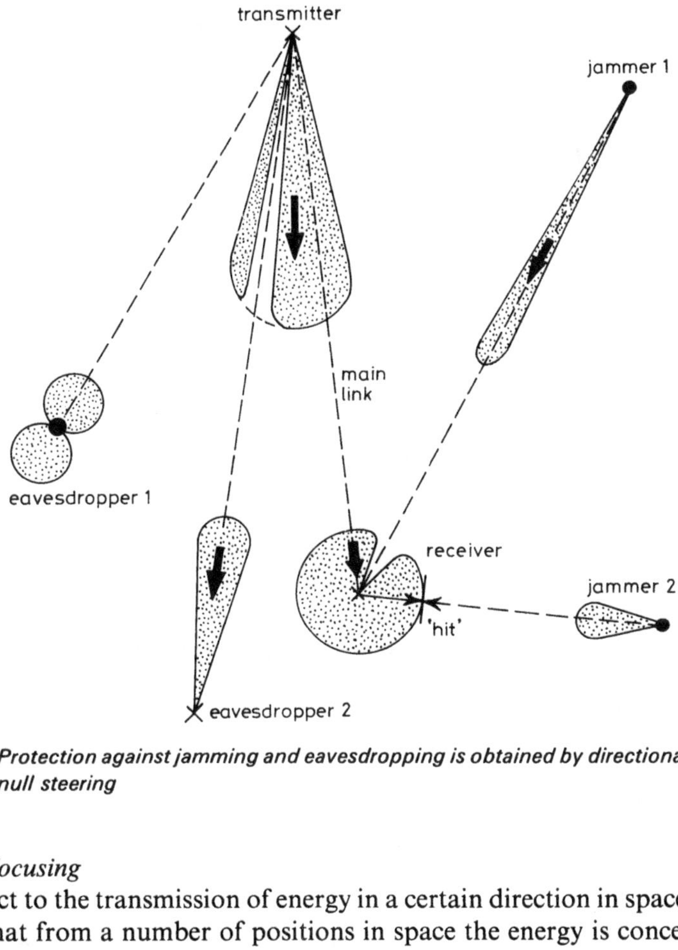

Fig. 2.18 *Protection against jamming and eavesdropping is obtained by directional beams and null steering*

2.5.1.2 *Focusing*

In contract to the transmission of energy in a certain direction in space, focusing implies that from a number of positions in space the energy is concentrated at one point.

The simplest form of focusing is the illumination of a point with several independent beams. If those beams are coherent, then the response at the illuminated spot will be proportional to the number of beams illuminating. Coherent beams require that the illuminators (transmitters or receivers) are synchronized. If they are not, then the bandwidth of the focusing signal waveform will be affected.

Focusing may be used to concentrate jamming power in certain areas of interest. Depending on the type of jamming, different requirements will form the mutual coherency of the jamming waveforms coming from the various sources.

Consider two sources A and B jamming point C. The distance from A to C is R_{AC}, and the distance from B to C is R_{BC}. The jamming waveform is $j(t)$. At point C, the contribution to the field strength from jammer A is $j_A(t)$ and from jammer B is $j_B(t)$. j_A and j_B have delays equal to R_{AC}/c and R_{BC}/c, respectively, from the point of transmission to point C. Neglecting the influence of propagation on signal amplitude, the signal at point C can be written as:

$$j_C(t) = j_A(t) + j_B(t) = j\left(t - \frac{R_{AC}}{c}\right) + j\left(t - \frac{R_{BC}}{c}\right)$$

The mean power of this signal is, assuming a stationary jamming signal:

$$\langle j_C^2(t)\rangle = 2\langle j^2(t)\rangle + \left\langle j\left(t - \frac{R_{AC}}{c}\right) j\left(t - \frac{R_{BC}}{c}\right)\right\rangle$$

The second term on the right-hand side is the autocorrelation of the signal for a lag equal to the difference in propagation time. Assuming the bandwidth of the jamming signal is B_j, the corresponding decorrelation time is

$$T_j = 1/B_j$$

and the second term contributes only if the time lag is less than the decorrelation time. Hence the following criterion holds:

$$\frac{R_{BC} - R_{AC}}{c} \ll \frac{1}{B_j}$$

For arbitrary time lags, each transmitter has to be synchronized to introduce initial time lags to compensate for the propagation time lags.From this follows the relationship between jamming station relative synchronization (BT_j) and jammer bandwidth B_j:

$$\Delta T_j \ll 1/B_j$$

This implies that a narrowband jammer requires no mutual synchronization, whereas a broadband jammer will require very strict time synchronization. Note also that the required synchronization accuracy is independent of jammer waveform.

If n stations (transmitters or receivers) are distributed in space, each having antenna gain $E_{antenna}$, then the space domain coding efficiency will be greatest when all the beams are directed at the same point (Fig. 2.19):

$$E_{space} = \begin{cases} 0 & \text{adjacent to the beams} \\ E_{antenna} & \text{in each antenna beam} \\ \sum_n E_{antenna} & \text{at the focal point} \end{cases}$$

This formula neglects the range dependence of field strength and assumes time synchronization to be sufficient for the bandwidth used (see earlier). If time synchronization is not used, then no coherent addition of field strength will occur and the E_{space} factor will be drastically reduced.

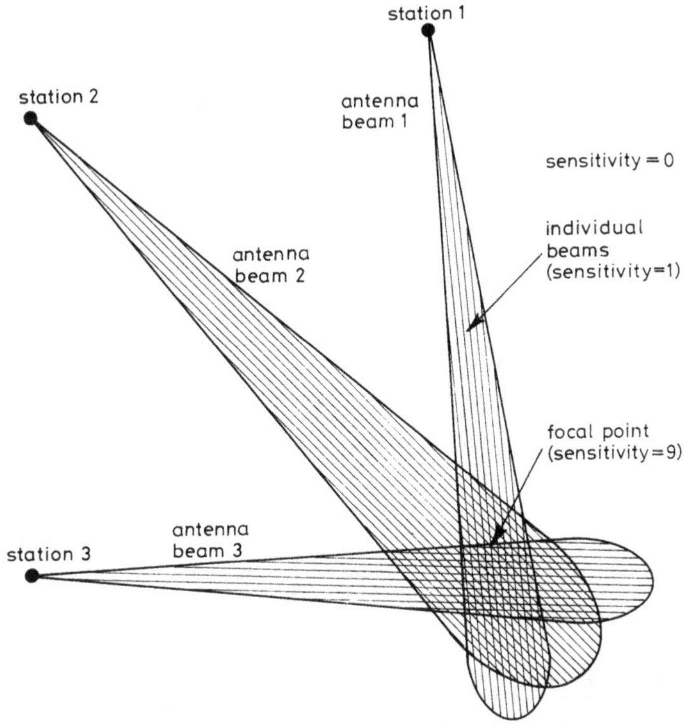

Fig. 2.19 *By illuminating a segment in space with several rays and compensating for the propagation time, coherent addition of energy is achieved. This can be used for concentrating of jamming power at the location of a receiver or for optimum detection of a transmitter conceived in a noisy environment*

The relative time lag between the stations in space is the only factor affecting the field strength in the focal point. The absolute time reference will be of importance when countermeasures requiring time of waveforms are used. As shown in Section 2.4, the exploitation of the time reference is crucial to many intelligent countermeasures. Therefore, the space domain efficiency factor has to include the effect of spatial variability of the time reference. By proper timing, focusing of signal power can be achieved and, in addition, the time of arrival can coincide with the synchronization of the attacked station.

Owing to the reciprocity of the electromagnetic waves, the same procedure can be applied equally well for reception. This can be useful to eliminate interference sources which could not be adequately suppressed by other means. Also, the position of transmitters can be found using these procedures. By measuring the direction of the phase front at two or more positions in space, the location of the transmitter can be established.

2.5.2 *Transverse properties of electromagnetic wave: polarization*
The electric field of an electromagnetic wave can be represented as:

$$E(f) = E_H(f)\hat{\boldsymbol{h}}_H + E_V(f)\hat{\boldsymbol{h}}_V$$

where $\hat{\boldsymbol{h}}_H$ and $\hat{\boldsymbol{h}}_V$ represent the orthogonal unit vectors in the horizontal and vertical plane. Fig. 2.20 indicates the transversal nature of the electromagnetic wave.

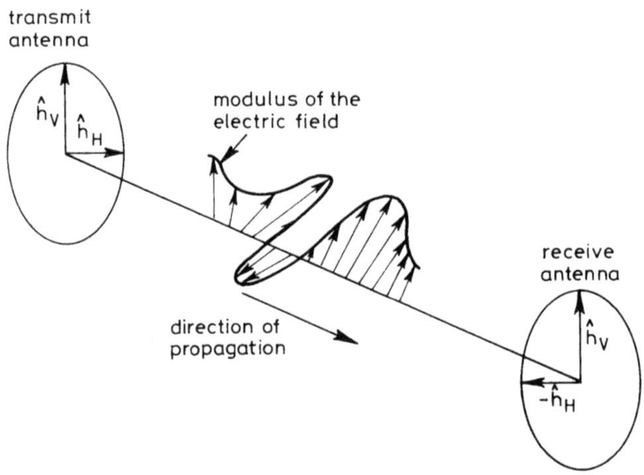

Fig. 2.20 *Polarization properties of the electromagnetic wave is described by the modulus and transversal direction of the electromagnetic field. For a narrowband (sinusoidal) signal, the trajectory is ellipsoidal*

It should at this stage be noted that as the modulus of the electric field vector is the vector sum of the horizontal and vertical field components, it can only be zero if *both* the components are zero. This implies that the transmitter can operate the two polarizations simultaneously, but it cannot conceal energy in one polarization channel by adding a random or deterministic signal in another. This rather obvious fact is of course a result of the independence of the two dimensions, but should be kept in mind when the polarization of the wave is applied for anti-eavesdropping purposes.

The polarization of an electromagnetic wave is determined by the polarization properties of the transmitting antenna. If the antenna transmits in a given direction a horizontal polarization with an efficiency η_H^T and a delay τ_H^T, and a vertical polarization with efficiency η_V^T and delay τ_V^T, the emitted field will be given as:

$$E_e(t) = V_T(t - \tau_H^T)\eta_H^T\hat{\boldsymbol{h}}_H + V_T(t - \tau_V^T)\eta_V^T\hat{\boldsymbol{h}}_V$$

$V_T(t)$ is the input to the antenna. Assume that this field hits a receiving antenna with efficiencies η_H^R and η_V^R and time lags τ_H^R and τ_V^R. The receiving antenna faces the opposite direction relative to the transmitting antenna; one of the unit vectors in the receiving space will have to change sign, as the antenna is always characterized in its transmitting mode. The output signal from the receiving antenna thus becomes:

$$V_R(t) = V_T(t - \tau_H^T - \tau_H^R)\eta_H^T\eta_H^R \boldsymbol{h}_H^T \boldsymbol{h}_H^R + V_T(t - \tau_V^T - \tau_V^R)\eta_V^T\eta_V^R \boldsymbol{h}_V^T \boldsymbol{h}_V^R$$

From this we see that in order to maximize the output of the receiver, we can choose one of several strategies. The simplest is that one polarization only can be received. This is a suboptimal strategy as the energy in the orthogonal polarization is not utilized, but it is advantageous when the reference axis is not known. In this case the relative time lags of the antenna polarization channels are of no importance when the reference axes are perfectly aligned, and the optimum strategy is simply to detect the component with the highest power.

The variance of the received signal is then given as

$$\begin{aligned} E[V_R^2(t)] &= E[V_T^2(t - \tau_H^T - \tau_H^R)(\eta_H^T\eta_H^R \boldsymbol{h}_H^T \boldsymbol{h}_H^R)^2] \\ &+ E[V_T^2(t - \tau_V^T - \tau_V^R)(\eta_V^T\eta_V^R \boldsymbol{h}_V^T \boldsymbol{h}_V^R)^2] \\ &+ 2E[V_T(t - \tau_H^T - \tau_H^R)V_T(t - \tau_V^T - \tau_V^R) \\ &\times \eta_H^T\eta_H^R\eta_V^T\eta_V^R \boldsymbol{h}_H^T \boldsymbol{h}_H^R \boldsymbol{h}_V^T \boldsymbol{h}_V^R] \end{aligned}$$

If V_T is a stationary process with zero mean and power P_T, the received power P_R is given by:

$$\begin{aligned} P_R &= E[V_R^2(t)] \\ &= P_T\{(\eta_H^T\eta_H^R)^2 + (\eta_V^T\eta_V^R)^2 \\ &+ \text{acf } V_T[(\tau_H^T - \tau_H^R) - (\tau_V^R - \tau_V^T)] \\ &\times \eta_H^T\eta_H^R\eta_V^T\eta_V^R \boldsymbol{h}_H^T \boldsymbol{h}_H^R \boldsymbol{h}_V^T \boldsymbol{h}_V^R\} \end{aligned}$$

acf $V_T(\Delta\tau)$ is the normalized autocorrelation function with lag $\Delta\tau$ at the modulation signal V_T.

From this equation it can be seen that the optimum strategy for the receiver (eavesdropper) is:

1. If either of η_H^T or η_V^T is zero, then the corresponding receiver η should also be zero. This means that if transmission is performed on one polarization only, all the available receiving antenna gain should be utilized on this polarization.
2. When both η_H^T and η_V^T are nonzero, two criteria should be met. First, to allow the acf term to be maximum (equal to 1), the time lag through both antenna channels should be equal:

$$\tau_H^T + \tau_H^R = \tau_V^R + \tau_V^T$$

Next, the reference axis should be aligned antipodally:

$$\hat{h}^{T}_{H}\hat{h}^{R}_{H}\hat{h}^{T}_{V}\hat{h}^{R}_{V} = +1$$

By using the polarization properties of the electromagnetic waves, a set of different countermeasure techniques can be devised. The two degrees of freedom offered by polarization are independent, and independent coding schemes along the two axes are feasible.

2.6 Case study: spread spectrum communication link in spatial scenario with jammer and eavesdropper

To illustrate how the various techniques give different characteristics, assume that a spread spectrum communication link is introduced in a scenario such as that of Fig. 2.21. The main transmitter A directs a beam towards receiver B. The jammer positioned at D tries to interfere with the receiver by transmitting a

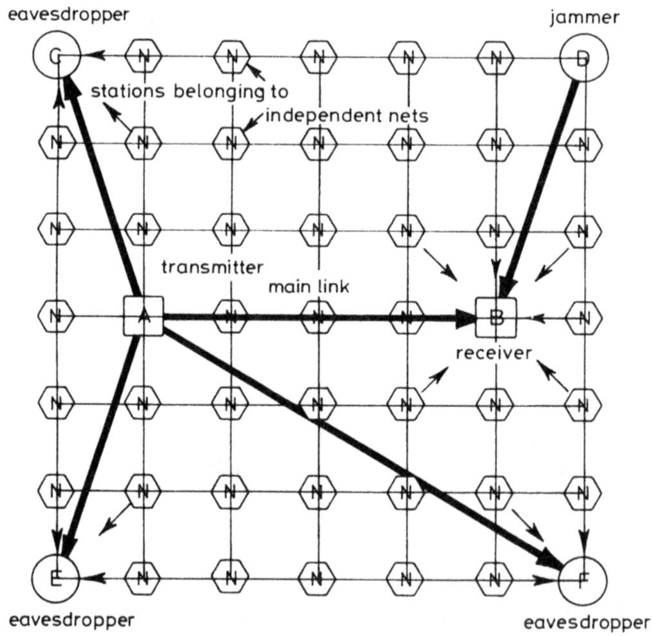

Fig. 2.21 Communication scenario with transmitter and receiver pair working in an environment with noise sources, jammers and eavesdroppers

jamming waveform. The eavesdropper at E listens to the transmitter from a single point in space, whereas eavesdroppers C and E are synchronized so as to provide spatial focusing at point A. A number of nets independent of the A–B links contribute a spatially distributed noise to the receiver and the eavesdropper.

To analyse this scenario, the interaction efficiency matrix of the stations in this net is expressed as shown in Fig. 2.22. Transmitter A manages to transfer information to receiver B with efficiency E_{AB}, and unintentionally to eavesdroppers C, E and F with efficiencies E_{AC}, E_{AE}, and E_{AF} respectively. The receiver is influenced by the jammer with efficiency E_{DB} and also by the noisy background with E_{NB}. The eavesdroppers are also similarly jammed by the jammer and the noise.

countermeasure efficiency from this station \ countermeasure efficiency to this station	A	B	C	D	E	F	N
A		E_{AB}	E_{AC}		E_{AE}	E_{AF}	
B	no transmission						
C	no transmission						
D		E_{DB}	E_{DC}		E_{DE}	E_{DF}	
E	no transmission						
F	no transmission						
N		E_{NB}	E_{NC}		E_{NE}	E_{NF}	

Fig. 2.22 *Interaction matrix of the net in Fig. 2.21*

The efficiency operator is, as introduced in Section 2.1, the total product of the efficiency of a number of independent operations. Following the analysis in Sections 2.4 and 2.5, the efficiency can be written as:

$$
\begin{aligned}
E = \; & E_{PG} && \text{processing gain factor} \\
& \times E_{\delta T} && \text{time uncertainty factor} \\
& \times E_{\delta F} && \text{frequency uncertainty factor} \\
& \times E_{\text{beam}} && \text{antenna response factor} \\
& \times E_{\text{pol}} && \text{polarization response factor} \\
& \times E_{\text{focus}} && \text{focusing response factor}
\end{aligned}
$$

Consider an example. The transmitter is a direct sequence system with processing gain $PG = 1000$. The code is known to jammer D and eavesdropper F but not to eavesdroppers C and E. This gives efficiencies 1000 or 1 for the E_{PG} element efficiency matrix. The transmitted power (field strength) is assumed to be 10. If furthermore only the receiver and eavesdropper F are synchronized, the $E_{\delta T}$ component of the matrix will be 1000 for E_{AB} and E_{AF}.

If it is assumed that the frequency band is known to everybody, the $E_{\delta F}$ contribution will be 1 for all the elements.

Furthermore, assume that the transmitter has an antenna highly directional towards the receiver, and the receiver has an omnidirectional antenna with a null in the direction of the jammer. Jammer D and eavesdropper F have omnidirectional antennas, and eavesdroppers C and E have directional antennas. The antenna beam matrix is then as in Fig. 2.23. The description ‘highly directional’ is given the score 1000, ‘directional’ the score 100 and ‘omnidirectional’ the score 1.

radiation from \ radiation to	A	B	C	D	E	F
A		1000	1	x	1	1
B	1		x	0	x	x
C	100	1		1	1	1
D	1	1	1		1	1
E	100	1	1	1		
F	1	1	1	1	1	
N	1	1	1	1	1	1

Fig. 2.23 *Antenna response matrix*

energy radiated from \ energy radiated to	A	B	C	D	E	F	N
A		10^{10}	2000		2000	0	
B							
C							
D		0	1		1	0	
E							
F							
N		1	1		1	0	

Fig. 2.24 *Total efficiency of operations in Fig. 2.21*

By taking the product of the antenna gain of the transmit and receive antennas for each combination of antennas, the E_{beam} factor can be calculated.

The polarization can be assumed to be horizontal, and the only station using vertical polarization is eavesdropper F.

Table 2.2

	E_{AB}	E_{AC}	E_{AE}	E_{AF}	E_{DB}	E_{DC}	E_{DE}	E_{DF}	E_{NB}	E_{NC}	E_{NE}	E_{NF}
E_0	10	10	10	10	1	1	1	1	1	1	1	1
E_{PG}	1000	1	1	1000	1000	1	1	1000	1	1	1	1
$E_{\delta T}$	1000	1	1	1000	1000	1	1	1	1	1	1	1
$E_{\delta F}$	1	1	1	1	1	1	1	1	1	1	1	1
E_{beam}	1000	100	100	1	0	1	1	1	1	1	1	1
E_{pol}	1	1	1	0	1	1	1	0	1	1	1	0
E_{focus}	1	2	2	1	1	1	1	1	1	1	1	1
Total	10^{10}	2000	2000	0	0	1	1	0	1	1	1	0

The focusing efficiency factor will only be valid for eavesdroppers C and E, which will have a focusing gain equal to 2 with respect to the transmitter A.

Assuming the energy (field strength) emitted by the jammer and the noise source is 1, the total efficiency can be calculated as in Table 2.2. No range dependencies of field strength are considered. These are easily calculated using the square root relationship.

Fig. 2.24 shows the total efficiency of the operations in matrix form. The success of an operation is checked by looking at the score of each countermeasure. It is seen that the score for the transmission process from A to B is 10^{10}, which is well above the score of 1 from the noise sources D and N. Eavesdropper F does not succeed and gets a score of 0. Eavesdroppers C and E score 2000 for reception from the transmitter; this is more than the score from the noise sources (1), which shows that these operations have been successful!

Chapter 3

Coding for bandwidth spreading

In direct sequence systems the spectrum of the information signal is expanded, using codes at the transmitter, to occupy the whole of the available bandwidth. In code division multiple access the channel separation is primarily due to coding. Thus the code sequences are very important to a spread spectrum system. The type of code used sets bounds on the capability of the system, bounds that can only be changed by changing the codes.

This chapter discusses some of the requirements for the spreading codes with respect to synchronization, multiple access systems and the spectral occupancy of the codes. The last is of special interest in a system to achieve concealment.

The spreading codes can be identified by a large number of parameters, but this discussion is restricted to a smaller set, important in any design and performance analysis. Unfortunately quantitative results are hard to compute for several of the important parameters, since general analytical expressions valid for a particular sequence of set of sequences may not exist.

In these cases evaluation of upper and lower bounds for a set of sequences is useful for calculating worst case, as well as average, performance. The bounds also indicate what may be gained by the effort of going through optimalization algorithms. Perhaps just as important for calculating and using bounds are the tradeoffs between maximum magnitudes of different important parameters.

3.1. Fundamental model

The transmitted signal $s^k(t)$ for some user k may be represented as

$$s^k(t) = A^k a^k(t) b^k(t) \cos(\omega_c t + \alpha^k) \tag{3.1}$$

where A^k is the signal amplitude. $b^k(t)$ is the kth user's data signal, such that

$$b^k(t) = \boldsymbol{b}^k = \ldots b_{-1}, b_0, b_1, \ldots$$

where b_i is a sequence of unit amplitude positive and negative pulses of duration T_d. The code sequence $a^k(t)$ used for bandwidth expansion is similarly given by

$$a^k(t) = \boldsymbol{a}^k = (a_0, a_1, \ldots, a_{p-1})$$

where a_i is $+1$ or -1 and p is the period of the sequence. α^k in eqn 3.1 represents the phase of the kth carrier signal and ω_c is the carrier frequency.

The actual transmitted sequence for a system using sequence inversion keying is thus

$$\ldots, b_{-1}\boldsymbol{a}^k, b_0\boldsymbol{a}^k, b_1\boldsymbol{a}^k, \ldots$$

The received signal for the case of one user $\boldsymbol{k}$ and with thermal noise is given by

$$r(t) = n(t) + A^k a^k(t - \tau^k) b_i^k(t - \tau^k) \cos[\omega_c t + (\alpha^k - \omega_c \tau^k)] \quad (3.2)$$

The matched filter receiver recovers the data by correlating with $a^k(t)\cos\omega_c t$ to give the output

$$\begin{aligned} Y^k &= \int_0^{T_d} r(t) a^k(t) \cos \omega_c t \, dt \\ &= \frac{1}{2} A^k b_1^k T_d + \int_0^{T_d} n(t) a^k(t) \cos \omega_c t \, dt \end{aligned} \quad (3.3)$$

at $t = T_d$, ignoring the double frequency component of $r(t)\cos\omega_c t$ and putting α^k and τ^k to zero owing to synchronization. If the matched filter is implemented digitally, the input will be sampled, usually after the carrier is removed (see Fig. 3.1.).

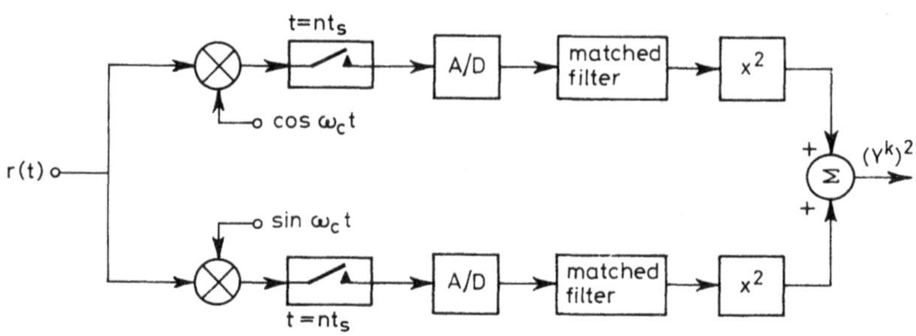

Fig. 3.1 *Lowpass version of a digital matched filter*

The lowpass implementation usually involves a quadrature technique because no restoration of carrier phase coherence is feasible. The input signal is first heterodyned by the use of two sinusoids at carrier frequency which are in phase quadrature. The signals are then passed into matched filters, the filter outputs are squared and added, and the square roots are taken to form the final output. However, if it is assumed that α^k and τ^k can be set to zero owing to synchronization, the analysis is restricted to the upper branch of Fig. 3.1. The digital filter impulse response is then derived by sampling the impulse response of the analogue filter. In this case Y^k can be written as

$$\begin{aligned} Y^k &= \sum_{l=0}^{p-1} r(lt_s) a_l^k \quad (3.4) \\ &= \tfrac{1}{2} A^k b_i^k p + \sum_{l=0}^{p-1} n(lt_s) a_l^k \end{aligned}$$

where $1/t_s$ is the sample rate; $n(lt_s)$ is the sampled lowpass version of $n(t)$. Eqn 3.4 will be used for the derivation of the different correlation parameters.

3.2. Autocorrelation parameters

When the receiver is l code elements (chips) out of phase with the data bit edges, as will be the case at least initially before the data fills the matched filter, the output is given by

$$\begin{aligned} Y^k(l) &= \tfrac{1}{2}A^k\{b_i(a_0^k, \ldots, a_{p-l-1}^k), [n(0), \ldots, n(p-1)]\}\, a^k \\ &= \tfrac{1}{2}A^k b_i C^k(l) + [n(0), \ldots, n(p-1)]a^k \end{aligned}$$

where

$$C^k(l) = \sum_{j=0}^{p-l-1} a_j^k\, a_{j+l}^k \qquad 0 < l < p \tag{3.5}$$

is the nonperiodic autocorrelation function. For any $l \neq 0$ it is desirable that $C(l)$ is small compared with p to avoid false synchronization.

For application over multipath channels, knowledge of the nonperiodic autocorrelation is also important. For a two-path model (Fig. 3.2) the correlation process may look like

$$\begin{aligned} Y^k(l) &= \tfrac{1}{2}A^k[b_i(a_0^k, \ldots, a_{p-1}^k) + b_{i-1}(a_{p-1}^k, \ldots, a_{p-1}^k) \\ &\quad + b_i(a_0^k, \ldots, a_{p-l-1}^k)]\boldsymbol{a}^k \\ &= \tfrac{1}{2}A^k[p + b_{i-1}C^k(p-l) + b_iC^k(l)] \end{aligned} \tag{3.6}$$

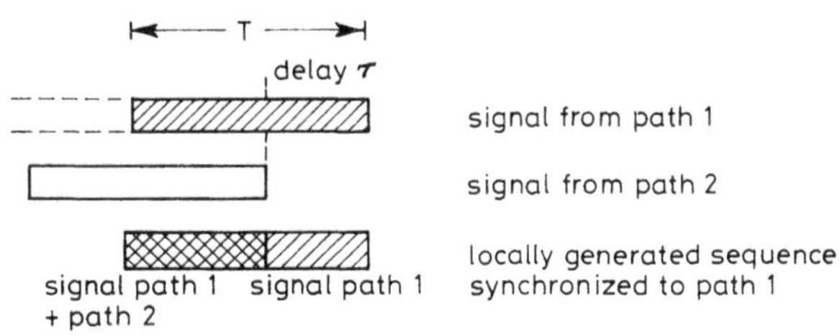

Fig. 3.2 *Multipath channel*

where $C^k(p-l)$ is defined from eqn 3.5 when l is replaced by $(p-l)$, so that

$$C^k(p-l) = \sum_{j=0}^{l-1} a_j^k\, a_{j+p-l}^k$$

The first term on the right-hand side of eqn 3.6 is the even (periodic) autocorrelation function resulting from the propagation path to which the receiver is aligned. The next two terms are due to the second path being delayed by lT_c sequence chips. There are two cases to be considered. If $b_{i-1} = b_i$ then the sum of the last two terms is also an even (periodic) autocorrelation function. Thus the even correlation function is related to the nonperiodic autocorrelation functions by

$$\theta^k = C^k(l) + C^k(p - l) \tag{3.7}$$

When $b_{i-1} \neq b_i$ the odd correlation $\hat{\theta}^k$ is given by:

$$\hat{\theta}^k = C^k(l) - C^k(p - l) \tag{3.8}$$

The performance of the spread spectrum system in a multipath situation will thus depend on the spreading code characterized by their even and odd autocorrelation properties.

The way in which codes should be optimized will depend on the performance criteria chosen for the system analysis. This discussion is however postponed until Section 3.4, where a system of K users in a multiple access system is treated. The performance analysis of the two-path propagatin model considered may in fact be carried out using the multiple access equations with $K = 2$ and with the crosscorrelation functions replaced by the corresponding autocorrelation functions.

3.3. Crosscorrelation parameters

If the spread spectrum system is expanded to contain K users at the same carrier frequency in a code division multiple access system, eqn 3.2 can be rewritten as

$$r(t) = n(t) + \sum_{r=1}^{K} A^r a^r(t - \tau^r) b_i^r(t - \tau^r) \cos [\omega_c t + (\alpha^r - \omega_c \tau^r)] \tag{3.9}$$

The signal may then be correlated with $a^k(t) \cos \omega t$, $k \in [K]$, which for the digital realization gives the output

$$Y^k = \tfrac{1}{2} A^k b_i^k p + \sum_{\substack{r=1 \\ r \neq k}}^{K} \tfrac{1}{2} A^r \{b_{i-1}^r [a_0^r, \ldots, a_{p-l(r)-1}^r] + b_i^r [a_{p-l(r)}^r, \ldots, a_{p-1}^r]\} \boldsymbol{a}^k \cos \beta^r + \sum_{l=0}^{p-1} n(lt_s) a_l^k \tag{3.10}$$

where $l(r) \in [0, 1, 2, \ldots, p - 1]$ and β^r is the phase angle between the kth carrier and carrier r. In proceeding from eqn 3.9 to eqn 3.10 it is assumed that $\alpha^k = 0$ and $\tau^k = 0$.

Making use of nonperiodic crosscorrelation functions $C^{kr}(l)$ and $C^{kr}(p - l)$, defined similarly to the nonperiodic autocorrelation, we get

$$Y^k = \tfrac{1}{2}A^k b_i^k p + \sum_{\substack{r=1 \\ r\neq k}}^{K} \tfrac{1}{2}A^r[b_{i-1}^r C^{kr}(l) + b_i^r C^{kr}(p - l)] \cos \beta^r + \sum_{l=0}^{p-1} n(lt_s)a_l^k \tag{3.11}$$

Again there are two cases to be considered for each of the $K - 1$ interferences influencing the kth user. If $b_{i-1}^r = b_i^r$ for some $r \in [K - 1]$, where $[K - 1]$ is identical to $[K]$ without the kth user, then the interference will depend on

$$\theta^{kr} = C^{kr}(l) + C^{kr}(p - l) \tag{3.12}$$

which is called the even or periodic crosscorrelation. If $b_{i-1}^r \neq b_i^r$ then the interference will depend on

$$\hat{\theta}^{kr} = C^{kr}(l) - C^{kr}(p - l) \tag{3.13}$$

which is called the odd crosscorrelation.

3.4. Performance of phase coded spread spectrum multiple access systems

The performance of the multiple access system will depend on the correlation properties of the set of spreading codes. The system use will determine on which properties attention should be focused. Ideally one would like to find codes for which the error probabilities are small under all possible propagation conditions and data transmissions.

Such an outcome may result in a minimax alternative where emphasis is put on finding codes which give the smallest value of the maximum error probability. Hence the maximum value of the error probability will be minimized by selecting a set of spreading codes for which the peak even and odd crosscorrelations are small for every possible $0 < l < p - 1$.

Pursley [1] derived the worst case of detection error as

$$P_{e\max} = 1 - \Phi\left\{\left[1 - \frac{\sum_{2}^{K} \max(\theta, \hat{\theta})}{p}\right]\sqrt{(2E/N_0)}\right\} \tag{3.14}$$

where $\theta = \max(|\theta^{kr}(l)| : 0 \leqslant l \leqslant p - 1)$ and $\hat{\theta} = \max(|\hat{\theta}^{kr}(l)| : 0 \leqslant l \leqslant p - 1)$, Φ is the standard Gaussian cumulative distribution function, E is the energy per data bit and $N_0/2$ is the two-sided spectral density of the noise. Eqn 3.14 provides an upper bound on the worst case error probability. However, for some combinations of K and p the term $[\sum_{2}^{K} \max(\theta, \hat{\theta})]/p$ may become greater than unity and eqn 3.14 will be of no value.

If the phase, time and data bit variables are taken to be mutually independent random variables, it is possible to calculate their probabilistic averages and thus important performance measures such as average signal/noise ratio and average error probability.

If the interference is treated as additional noise, the average signal/noise ratio can be defined as [1]

$$SNR = [E(Y^k | b_0^k)]/[\text{var}\,(Y^k | b_0^k)] \tag{3.15}$$

where it is measured that the data bit b_n^k is a random variable taking values $+1$ and -1 with equal probability.

The signal/noise ratio for a matched filter at the instant of sampling has been derived for white noise to be [2]

$$SNR = (2E/N_0)^{1/2}$$

E is the signal energy and $N_0/2$ is the two-sided spectral density of the noise. The variance of the noise due to code interference is derived in [3] and is given by

$$\sigma_I^2 = \frac{1}{24p}(A^k)^2 \sum_{\substack{k=1 \\ k \neq r}}^{K} S^{kr} \tag{3.16}$$

where S^{kr} is given by

$$S^{kr} = \sum_{l=0}^{p-1} [C^{kr}(l)]^2 + C^{kr}(l)C^{kr}(l+1) + [C^{kr}(l+1)]^2 + [C^{kr}(l-p)]^2$$
$$+ C^{kr}(l-p)C^{kr}(l-p+1) + [C^{kr}(l-p+1)]^2 \tag{3.17}$$

In eqn 3.17 any restriction on the time delays to integer code chip values is removed in order to allow calculation of average signal/noise ratios. The terms $C^{kr}(l+1)$ and $C^{kr}(l-p+1)$ are introduced for this reason, assuming that the C^{kr} are linear continuous functions on the interval $[lT_c, (l+1)T_c]$. Eqn 3.16 is thus computed by means of probabilistic expectations with respect to carrier phases, time delays and data bits.

The final *SNR* for a system with white noise and code interference then becomes

$$SNR = \left[\frac{E(Y^k | b_0^k)^2}{(\sigma_I^2 + \sigma_n^2)}\right]^{1/2} = \left[\frac{\sigma_I^2}{E(Y^k | b_0^k)^2} + \frac{\sigma_n^2}{E(Y^k | b_0^k)^2}\right]^{-1/2}$$

where σ_n and σ_I refer to the white noise process and the code interference respectively. The processes are assumed to be independent, and so they can be added on a power basis.

Now $E(Y^k | b_0^k) = \frac{1}{2}A^k p$, and the SNR is given by

$$SNR = \left[\frac{\frac{1}{24p}(A^k)^2 \sum_{\substack{r=1 \\ r \neq k}}^{K} S^{kr}}{(\frac{1}{2}A^k p)^2} + \frac{N_0}{2E} \right]^{-1/2}$$

$$= \left(\frac{1}{6p^3} \sum_{\substack{r=1 \\ r \neq k}}^{K} S^{kr} + \frac{N_0}{2E} \right)^{-1/2}$$

$$= \left(Q^{kr} + \frac{N_0}{2E} \right)^{-1/2} \tag{3.18}$$

Q^{kr} defines the interference due to the other users of the multiple access system, and from eqn 3.17 it will depend on the periodic and nonperiodic crosscorrelation functions. The evaluation of eqn 3.18 involves a dramatically large amount of computation. If there are K users there is a need to calculate (2^K) crosscorrelation functions and S^{kr}. However, a proposition by Gold [4] can be used to reduce the computation to only K autocorrelation functions. In Appendix A it is shown that

$$\sum_{l=0}^{p-1} [C^{kr}(l-p)]^2 + [C^{kr}(l-p+1)]^2 + [C^{kr}(l)]^2 + [C^{kr}(l+1)]^2$$

$$= 2p^2 + 4 \sum_{l=1}^{p-1} C^k(l)\, C^r(l)$$

Neglecting the terms $C^{kr}(l)C^{kr}(l+1)$ and $C^{kr}(l-p)C^{kr}(l-p+1)$ has been shown to be quite satisfactory even for moderate values of p. S^{kr} can then be simplified, and eqn 3.18 becomes

$$SNR = \left\{ \frac{1}{6p^3} \sum_{\substack{r=1 \\ r \neq k}}^{K} \left[2p^2 + 4 \sum_{l=1}^{p-1} C^k(l) C^r(l) + \frac{N_0}{2E} \right] \right\}^{-1/2} \tag{3.19}$$

The importance of the last simplification lies in the fact that the SNR may now be calculated using only K autocorrelation functions. Yao [5] showed that $P_e = 1 - \Phi(\sqrt{SNR})$, where SNR is defined as in eqn 3.18, is a very good approximation of the actual average error probability for many practical values of p and K. Yao [5] also gives upper and lower bounds on the error probability applying moment space bounds. For application of these bounds the maximum periodic and odd crosscorrelation must be computed.

So far a system using sequence inversion keying has been assumed for data modulation. Another scheme often applied is binary orthogonal keying, which uses different sequences for the $+1$ and -1 data bits. This requires knowledge of the crosscorrelation functions between $2K$ sequences. In fact the changes in S^{kr} will be the replacement of $C^{kr}(l-p)$ and $C^{kr}(l-p+1)$ by $C^{k'r}(l-p)$ and $C^{k'r}(l-p+1)$, where k' denotes the sequence used for a data bit $= -1$.

One particular change in the latter system is the fact that the sequences used for $+1$ and -1 will interfere in the matched filter receiver to a degree depending on their mutual periodic crosscorrelation.

3.5. Signal concealment

If a transmission has a low power spectral density compared with thermal and environmental noise and if the spectrum is nearly flat, then it is difficult to distinguish on a spectrum analyser. The result is a certain concealment of the transmission, important in several military applications. For a periodic spreading code the result is a line spectrum where the lines are separated by $1/pT_c$. The data bits will tend to be nearly random in character so that it is plausible to model the data message as a random binary sequence. The probability for a $+1$ or a -1 is the same and the process has a mean value of zero. The power spectrum will then look like a $\sin^2 x/x^2$ function, as plotted in Fig. 3.3. The data bits are then multiplied (in the binary case, modulo 2 added) with the spreading code. If the data bandwidth $\beta \geqslant 2/pT_c$ the resulting spectrum for a spreading code giving a $\sin^2 x/x^2$ envelope will be approximately flat, as shown in Fig. 3.4. Since the data bandwidth $\beta \cong 2/T_d$, a condition for communication concealment from spectrum analysis is

$$pT_c \geqslant T_d$$

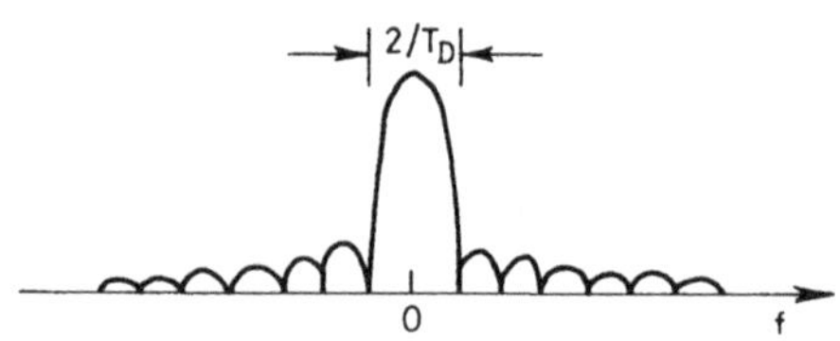

Fig. 3.3 *Frequency spectrum of data signal*

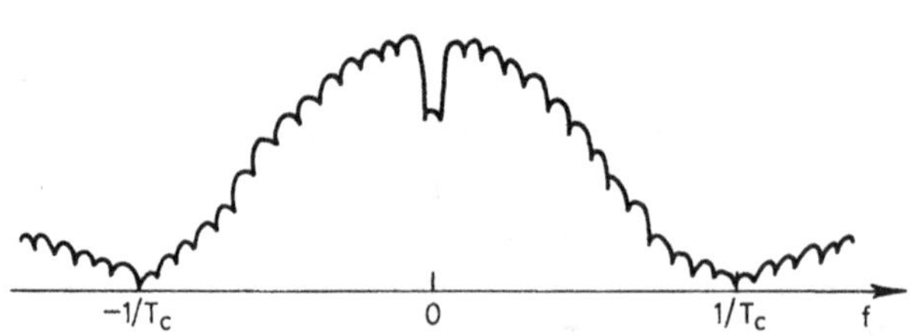

Fig. 3.4 *Frequency spectrum of data signal after addition of spreading sequence*

This means that each data bit should last for at most one period of the spreading code sequence.

The envelope of the continuous spectrum shown in Fig. 3.4 will be the same as for the spreading code. In order to keep the signal 'buried' in noise it is also important that the power spectrum is flat in the sense that most of the power is not restricted to only a few spectral components. This will depend on the actual runs of $+1$s and -1s in the spreading code.

3.6. Binary sequences for direct sequence spread spectrum systems

So far in the consideration of the effect of the spreading codes on the overall system performance, nothing has been assumed about the sequences other than a periodic property. The requirement for periodicity is a practical one, since a code that can be reproduced and aligned so that any transmitter–receiver time uncertainty is removed during synchronization must have a finite length.

However, when the receiver and transmitter are aligned and for as long as the clocks and control circuits keep the system synchronized, nonrepeated codes could be used. If, however, synchronization was lost it would be very difficult, if not impossible, to align the transmitter–receiver during data transmissions. Thus in burst transmissions, where the messages are fairly short, the data format usually contains a synchronization preamble; that is a periodic sequence followed by data bits which in some cases may be modulated by a long nonrepeated spreading code [6]. In this case synchronization takes place at the beginning of each message transfer.

The examination of possible burst spread spectrum systems in [6] revealed several correlation requirements for the spreading codes.

In order to perform the necessary synchronization of transmitter and receiver the codes needs an impulsive cyclic autocorrelation function. In order not to increase the possibilities of false lock and to lessen possible multipath interference the sidelobes described by the nonperiodic autocorrelation function should have small values.

In a system using mark and space spreading codes for signalling, the periodic crosscorrelation between the codes should be small. In a larger system operated as a code division multiple access network the performance is closely linked to both the periodic and nonperiodic crosscorrelation functions. If concealment of the transmission is important, the spreading code should produce a uniform spectrum of harmonics of reasonable constant amplitude up to some harmonic number.

It turns out that there exist families of possible spreading codes which are easy to generate and which satisfy some of these requirements very well, but for which it is necessary to apply a considerable computational effort in order to obtain subfamilies satisfying all the requirements reasonably well.

In the following, all binary sequences will be regarded as sequences of the 1 and the 0 for algebraic purposes, but as sequences of the -1 and the $+1$ when correlation functions are to be calculated. This is done by mapping the additive group with 1 and 0 to the multiplicative group with -1 and $+1$.

3.6.1. *Periodic autocorrelation of maximal-length sequences*
Maximal length shift register sequences (m-sequences) are certain binary sequences of length $p = 2^m - 1$, where m is the number of shift register stages.

A possible representation of the autocorrelation function is

$$\theta(i) = \frac{A - D}{p} = \frac{p - 2D}{p} \tag{3.20}$$

where A is the number of places that the code $a_0, \ldots, a_{p-1}$ with period p and the cyclic shift $a_i, \ldots, a_{i-1}$ agree, and D is the number of places where they disagree, so that $A + D = p$.

An m-sequence has the property that a period of the sequence contains 2^{m-1} 1s and 2^{m-1}-1 0s because there are 2^{m-1} odd numbers between 1 and 2^m-1 with binary representation ending in 1, and 2^{m-1}-1 even numbers in the same range with binary representation ending in 0.

The state of the shift register runs through these numbers since a window of width m slid along an m-sequence will see every 2^m-1 nonzero binary m-tuple once.

Then if we apply the property that the sum of an m-sequence and a cyclic shift of itself is another m-sequence, so that $\boldsymbol{a} + \boldsymbol{a}_i = \boldsymbol{a}_j$, eqn 3.20 becomes

$$\theta(i) = \frac{(2^{m-1} - 1)(2^{m-1})}{p} \qquad 0 < i < p - 1$$

When $i = 0$ then $\boldsymbol{a} = \boldsymbol{a}_i$ and $D = 0$, giving $\theta(0) = p/p = 1$.

Then alternatively the autocorrelation function for m-sequences is defined by

$$\begin{aligned} \theta(0) &= 1 \qquad i = 0 \\ \theta(i) &= 1/p \qquad 1 < i < p - 1 \end{aligned} \tag{3.21}$$

where the periodic normalized autocorrelation function $\theta(i)$ is defined by

$$\theta(i) = \frac{1}{p}\sum_{l=0}^{p-1} a_l a_{l+i}$$

For synchronization purposes this periodic autocorrelation function is ideal. In fact it can be shown [7] that this is the best possible autocorrelation function of any binary sequence of length 2^m-1 in the sense of minimizing $\max\theta(i)$, $0 < i < p$.

The final property justifying the name pseudorandom for m-sequences is the particular distribution of runs of 1s and 0s. A run is defined to be a maximal string of consecutive identical symbols. In any m-sequence, one half of the runs have length 1, one quarter have length 2, one eighth have length 3 and so on, as long as these fractions give an integral number of runs. In each case the number of runs of 0s is equal to the number of runs of 1s.

All of these properties are the same as one would expect from a coin tossing sequence. However, one way in which an m-sequence is seen to be not truly random is that the properties hold for every sequence, whereas in a coin tossing sequence there would be some variations from sequence to sequence.

The *m*-sequences have a concise description by polynomials. In the general case the output of an *m*-stage binary feedback shift register will satisfy a recurrence relation of order *m* (see Fig. 3.5):

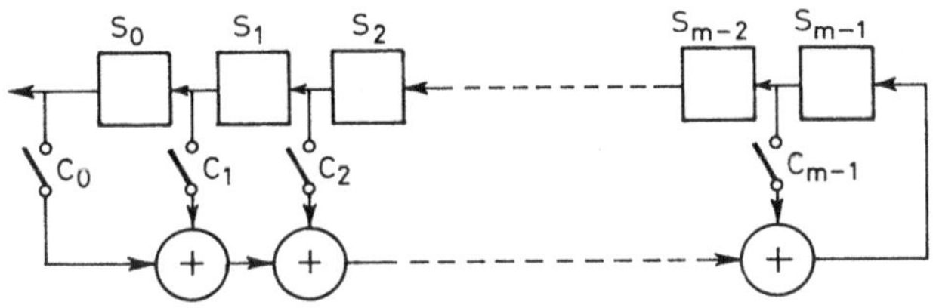

Fig. 3.5 *Binary feedback shift register*

$$S_{t+m} + C_{m-1}S_{t+m-1} + \ldots + C_1S_{t+1} + C_0S_t = 0 \tag{3.22}$$

The coefficients $C_i = 1$ if the stage S_i is added to the feedback path and $C_i = 0$ otherwise. The highest term S_{t+m} must always be present, and it is also assumed that $C_0 = 1$.

The solution to eqn 3.22 is closely related to the roots of the characteristic polynomial taking the form

$$f(x) = x^m + C_{m-1}x^{m-1} + \ldots + C_1x + C_0 \tag{3.23}$$

The number of such polynomials describing the feedback required to generate *m*-sequences is the same as the number of irreducible and primitive polynomials of degree *m*. This is given by Euler's phi functions:

$$\text{number of } m\text{-sequences of length } 2^m\text{-}1 = \frac{\phi(2^m\text{-}1)}{m} \tag{3.24}$$

Example 3.1
Consider a five-stage shift register which will produce *m*-sequences with periods of 31. From eqn 3.24 the number of such sequences is

$$\frac{\phi(2^5 - 1)}{5} = \frac{30}{5} = 6$$

The six m-sequences of period 31 will all have the ideal periodic autocorrelation described by eqn 3.21, and the periodic autocorrelation function will not depend on the initial content of the shift register. This is expressed by saying that the periodic autocorrelation function of *m*-sequences are insensitive to the phase of the sequence.

3.6.2. *Nonperiodic autocorrelation of maximal length sequences*
To complete the description of autocorrelation properties of *m*-sequences, calculation of the nonperiodic function is also needed.

The nonperiodic autocorrelation and thus the odd autocorrelation depend strongly on the phase of the sequences. That is to say, sidelobes and possible multipath interference will depend on the initial starting condition of the shift registers producing the spreading codes. This gives a multitude of codes for any given sequence period p corresponding to the p possible choices of phase for each of the cyclically distinct m-sequences. The result is roughly 10^6 possible codes of periods 31 to 4095 (there are 474 m-sequences of periods 31 to 4095), and there is a need for some criteria for sorting out the 'best' codes from this large selection. However, since no expression similar to eqn 3.21 exists for the nonperiodic parameters, one is left with an exhaustive search. Using a computer to find the phase for which $\hat{\theta}^k(l)$ is minimized will give, for several of the sequences, more than one possible phase. Hence such a first sieve does not generally give a unique set of optimal sequences.

A second condition that might be applied to reduce the probability of acquiring false synchronization, and which reduces the number of ties considerably, is to pick codes where $\hat{\theta}(l)$ attains its maximum value the least possible number of times. Both these sieves are concerned with peak odd autocorrelation values. However, in a multiple access system the degree of mutual interference is very important, as shown in Section 3.4. In particular, knowledge of the sidelobe energy of a sequence defined as

$$S(a^k) = \sum_{l=1}^{p-1} [C^k(l)]^2$$

can be used to bound the value of the average other user interference between codes [8]. This parameter can be selected as the final sieve in sorting out a cyclic shift optimal to the sequence in question.

Applying these sieves in successive order gives a unique cyclic shift for all the m-sequences with periods from 31 to 4095.

The conditions satisfied for optimal sequences with least sidelobe energy can then formally be written as;

(a) $\hat{\theta}^k(l)_{\max} \leqslant \hat{\theta}^k(T^s l) \qquad \forall\, s$

(b) When the number of elements of a finite set A is denoted by the cardinal number $|A|$, the cardinality of the set

$$\{l : |\hat{\theta}^k(l)| = \hat{\theta}^k_{\max},\ 0 < l \leqslant p-1\}$$

should be minimized.

(c) S^k is kept at a minimum.

Table 3.1 gives the numerical results of the codes chosen from these criteria for all m-sequences with periods from 31 to 4095 [9]. The first column identifies the m-sequence in octal notation [10]. The next column shows the initial or start position of the sequence when in auto-optimal phase. If the binary sequence $\boldsymbol{a}^k$ is generated by an m-stage feedback shift register, the initial shift register loading is given by a binary word of length $m = (a_0, a_1, \ldots, a_{m-1})$.

The third and fourth columns contain the same information as the first two for the reciprocal sequence. Notice that reciprocal pairs have identical nonperiodic autocorrelation values. The fifth column gives the peak odd autocorrelation for the m-sequence pair, and for comparison the peak value for the sequence in a 'worst' phase is shown in parentheses. The sixth column gives the number of times the peak correlation value occurs, and the final column gives the value of the sidelobe energy.

For many applications the sidelobe energy parameter may be more important than the peak correlation parameters. A reversal of the priority on peak correlation parameters and sidelobe energy, to give a set of phases which are auto-optimal among the set of least sidelobe energy phases, produces the numerical values in Table 3.2 [11]. This table is arranged similarly to Table 3.1 but no 'worst' phase peak values are given. This table only covers sequence periods from 31 to 255. Notice that there are a few sequences that are optimal irrespective of where the priority is put.

The periodic autocorrelation for m-sequences has been shown to be ideal. The nonperiodic autocorrelation parameters are obtained by a computer search which does not reveal any ideal m-sequences, only 'best' ones selected from a larger set. It would therefore be worth while to know how 'optimal' these m-sequences are compared with a more general set of possible periodic spreading codes. Sarwate [12] showed that the nonperiodic autocorrelation functions for a set of K complex valued sequences of period p are bounded by

$$\{\hat{\theta}_{\max}(l)\} \geqslant p\left[\frac{K-1}{pK-1}\right]^{1/2} \tag{3.25}$$

where

$$\hat{\theta}_{\max}(l) = \max|\hat{\theta}^k(l), \hat{\theta}^{kr}(l)|$$

For ranges of K from 2 to $K \approx p$ and large p, this lower bound varies from $\approx \sqrt{(\frac{1}{2}p)}$ to $\approx \sqrt{p}$. With $p = 1023$ the value for $\hat{\theta}_{\max}(l)$ using the bound in eqn 3.25 ranges from ≈ 23 to ≈ 32 when $K = 2$ and K is large respectively. In Table 3.1, $\hat{\theta}^k(l) > 57$, which is higher than the bound for $K = 2$ by more than a factor of two.

Golay [13] derives a 'merit factor' for general binary sequences as

$$F = \frac{p^2}{2S(\boldsymbol{a}^k)} \tag{3.26}$$

and establishes a conjectured asymptotic value for F that will be valid for very long binary sequences. The bound on F is given by

$$F_{\text{opt}} = \frac{12.3248}{(8\pi p)^{3/2p}} \tag{3.27}$$

From eqn 3.26 it is clear that choosing m-sequences with a high merit factor is the same as selecting the sequences by the least sidelobe energy criteria, the result of which is shown in Table 3.2. Calculating the merit factor for the sequences with smallest sidelobe energy from Table 3.2 and comparing them

Table 3.1 *Auto-optimal maximal length sequences with least sidelobe energy with periods from 31 to 4095*

	Sequence	Loading	Sequence	Loading	Min (max)	C	SE
$N = 5, p = 31$	45	11001	51	01001	7(11)	2	107
	67	00011	73	01101	7(13)	2	123
	75	11110	57	10010	7(13)	2	91
$N = 6, p = 63$	103	000010	141	011111	11(21)	2	427
	133	110001	155	011001	11(19)	2	503
	147	100011	163	110101	11(19)	2	435
$N = 7, p = 127$	301	0010010	203	1101101	17(29)	4	2087
	375	0101010	277	1110001	19(29)	4	2295
	367	0110101	357	1110010	17(27)	4	2563
	221	1001101	211	0010000	17(25)	6	2183
	361	1111111	217	0000101	15(29)	12	2015
	271	1000101	235	0001100	17(27)	10	2283
	345	0110001	247	0010111	17(25)	8	2255
	325	0000101	253	1101100	19(31)	6	2483
	323	1110111	313	1000111	17(25)	5	2203
$N = 8, p = 255$	561	11111110	435	10000101	25(37)	6	9199
	651	00011110	453	01000010	27(41)	2	10927
	551	00000111	455	01101111	27(43)	2	9607
	545	01001100	515	11011001	27(41)	4	9555
	765	00111100	537	10110100	25(37)	8	8883
	747	10000100	717	00111110	27(41)	2	10211
	615	01100010	543	10110001	27(39)	2	9875
	607	00100111	703	00111101	27(39)	4	9323

$N = 9, p = 511$	1041	010101001	1021	110000110	41(51)	2	43071
	1541	101011011	1033	000100110	39(53)	6	41319
	1321	111110011	1055	011011001	39(57)	6	39667
	1461	010001010	1063	110010010	41(61)	8	42599
	1151	011101010	1131	010101100	41(55)	4	42047
	1751	011101100	1137	100110010	43(63)	2	41143
	1731	101001010	1157	110011101	41(59)	14	42639
	1671	000010011	1167	111011010	39(51)	2	40683
	1371	101011000	1175	111010100	41(57)	2	37379
	1605	000001001	1207	110000010	43(61)	4	44943
	1245	101001010	1225	011010010	41(53)	2	43727
	1425	100111101	1243	000100001	45(67)	4	38567
	1725	010110010	1257	100000001	39(61)	12	43251
	1665	000110010	1267	100010111	43(59)	2	40691
	1365	001000001	1275	110000100	39(59)	6	36639
	1715	000110001	1317	111000100	39(49)	8	41315
	1555	100000000	1333	011111111	39(55)	2	36231
	1453	001000010	1423	101111011	39(53)	2	43475
	1743	000111110	1437	011001000	37(53)	6	43899
	1563	000000010	1473	100010111	45(63)	2	40703
	1713	001011111	1517	110010011	39(63)	8	42067
	1553	100110110	1533	110001110	39(55)	12	43903
	1773	100011110	1577	111111001	39(53)	2	38783
	1707	101010110	1617	110110010	43(59)	4	42035
$N = 10, p = 1023$	2201	0111011010	2011	1000011010	65(89)	4	180619
	3301	1110010110	2033	0110100100	59(85)	2	163255

Table 3.1 *Continued*

Sequence	Loading	Sequence	Loading	Min (max)	*C*	*SE*
3441	0101011111	2047	1011000011	61(79)	2	172447
2641	0100110010	2055	1000111101	67(91)	12	183191
2461	0000100101	2145	1011101011	63(87)	2	168055
3661	1110101010	2157	1100010000	63(77)	2	173907
3251	1001000110	2213	1000010100	63(79)	4	166347
2431	0110011111	2305	1010000100	61(83)	10	173135
3531	0000111111	2327	0011000111	61(79)	2	150899
3471	0011011101	2347	0000000100	61(79)	4	168043
3171	1011000100	2363	0100011001	65(83)	8	174739
3771	1101011111	2377	0001010101	65(89)	4	171311
2605	0000011000	2415	1001111000	59(79)	6	174751
3045	0101100111	2443	0110000010	63(91)	8	173651
2745	1101111010	2475	0010110100	65(85)	4	179551
3025	0111000000	2503	0111110001	57(75)	6	160751
3525	0011100000	2527	1000110000	51(79)	2	165495
3265	1000101100	2553	0010111010	59(77)	8	176443
3615	0110010011	2617	0110010111	57(81)	8	173039
3515	1011111100	2627	1001111000	61(79)	4	171771
3435	1000001001	2707	1011000000	61(85)	2	160551
3575	0001110001	2767	0000100110	59(79)	2	165445
3375	1111101011	2773	1010001000	59(73)	2	171959
3103	1011110110	3023	1000010100	57(73)	6	177079
3543	0000011010	3067	0000111110	57(71)	14	161507
3623	0100000000	3117	0011100001	61(75)	2	175635
3323	1000000000	3133	1110110000	59(85)	6	156639

	3763	0010010100	3177	0100000010	61(77)	2	167771
	3733	1000100000	3337	1001110000	59(75)	6	171395
	3507	0110011101	3427	0010101101	63(77)	2	168727
$N = 11, p = 2047$	5001	11000001000	4005	01010001110	87(109)	4	679399
	7201	11001101100	4027	11100100101	89(113)	10	724775
	6501	00100111111	4053	01110000001	89(127)	4	655027
	5501	11111100100	4055	01111010110	87(109)	2	645599
	7041	10010001000	4107	00111110000	97(121)	2	671363
	6141	01011010101	4143	01000110111	95(129)	4	685531
	5141	11011011101	4145	11100110001	99(135)	6	705091
	4341	11101111110	4161	11110001010	99(125)	6	698487
	6741	00000100010	4173	00101101101	85(111)	2	683707
	5421	01001001111	4215	10000001100	87(105)	4	702303
	5221	00111011100	4225	10100110111	87(109)	2	684415
	7621	00101100101	4237	01101011101	83(103)	4	687135
	4521	01000110111	4251	00100100101	97(123)	2	669359
	4321	11000011001	4261	00100000101	91(113)	4	647159
	7461	10010010000	4317	10100001001	95(123)	4	678159
	7161	10101010101	4347	11110010010	93(117)	2	684983
	6561	11001010010	4353	01101101110	93(113)	2	692399
	5361	00111001001	4365	10111000000	85(101)	8	706807
	5411	11100011010	4415	00011111100	95(121)	6	698179
	6211	11010011111	4423	11100000010	95(127)	4	685907
	5111	01000001011	4445	11000010011	91(121)	2	695799
	4511	11010001000	4451	11001100000	93(123)	2	656699
	6711	01110110010	4473	00011111111	89(109)	2	714279

Table 3.1 *Continued*

Sequence	Loading	Sequence	Loading	Min (max)	*C*	*SE*
5711	01010010110	4475	01110101100	93(111)	2	651655
5051	11101111100	4505	10001000011	95(123)	2	717475
6651	01110111001	4533	00100001000	91(127)	2	671163
6351	01100011001	4563	11010011011	93(119)	4	689767
5351	10000101010	4565	11110001010	85(109)	2	646827
7751	01010011010	4577	11110011101	95(117)	8	716807
6031	01100110000	4603	10101010100	87(111)	4	684951
7431	11010000011	4617	10000101001	87(105)	4	679407
6531	01001101100	4653	01001100111	89(117)	4	677883
5531	01111000100	4655	11111001000	87(107)	10	715547
4731	10001010010	4671	11011001100	83(107)	6	667623
7071	10001110000	4707	10001111111	95(119)	4	686439
5171	11111101101	4745	01110111101	89(111)	6	703859
7371	00100010100	4767	10111001000	93(129)	2	670651
7005	01010100010	5007	01001000110	93(117)	8	703967
6205	00011111101	5023	00111110110	87(119)	4	659467
5205	00101001111	5025	00010101110	91(119)	2	669727
5545	10001110111	5155	00100010101	89(113)	6	695047
7745	00111111101	5177	00100011100	103(127)	2	693607
5625	10111101101	5235	00110001001	103(127)	14	721647
7125	00011010111	5247	01101101000	89(109)	4	683831
6525	10010110011	5253	01011000111	99(123)	2	680683
6325	01101011111	5263	00011101101	85(107)	4	675643
5325	01110100100	5265	00111011001	85(109)	2	650607
7665	00000001101	5337	10110011111	93(119)	2	708311

7565	11010011111	5357	00101000011	89(119)	6	661571
6765	10000001100	5373	10101001101	93(117)	4	717027
6015	01101000000	5403	01011010101	95(125)	6	688567
6315	00101100010	5463	11011101101	91(113)	2	679979
7715	00001010110	5477	10010011001	99(125)	8	689423
6455	11111100000	5513	11000001100	91(111)	2	676983
7655	00110011000	5537	11010001010	87(109)	2	691239
7555	01101001111	5557	11001101011	87(117)	2	694755
5755	01110100110	5575	00001011000	89(111)	4	700931
7035	10011010011	5607	10110000101	91(117)	2	668443
6435	01001111111	5613	00010110110	99(119)	2	678387
6235	11011000101	5623	11101011010	93(115)	6	702543
7535	10110110101	5657	10010101011	91(117)	4	687455
7335	10001001101	5667	00001011111	85(111)	2	689255
5735	01111000100	5675	01001000101	97(123)	10	706607
6675	00011111000	5733	01111101010	83(107)	8	698039
7175	11010001101	5747	01100010011	89(113)	2	680315
6403	00010101001	6013	00011010000	87(105)	4	676007
7603	11000110010	6037	10000001011	87(107)	4	666235
7243	00100000011	6127	01001110011	83(105)	4	714191
6543	01110011011	6153	10110010111	83(101)	2	681847
6343	01100100110	6163	01110010101	93(117)	4	700903
7223	01100101100	6227	11001010001	95(117)	6	664463
6623	00011000100	6233	10000111011	91(113)	2	695203
6323	10001101111	6263	00110101000	95(123)	2	673435
7723	11101001011	6277	00110011111	99(121)	4	703419
7063	01110101110	6307	10011000010	97(119)	2	690927

Table 3.1 *Continued*

	Sequence	Loading	Sequence	Loading	Min (max)	C	SE
	7363	00111001001	6367	11101110001	85(107)	2	641831
	7413	01110111111	6417	10001011011	93(115)	2	684647
	7113	01010110110	6447	10101001110	99(133)	4	680099
	7053	11111011110	6507	10110000011	91(111)	4	694435
	7553	10111111111	6557	00101101110	99(115)	8	711223
	7633	01001000101	6637	00101011101	89(115)	2	682055
	6733	01101100001	6673	11010111011	91(109)	6	669379
	7273	01001100101	6727	10101000001	99(129)	4	699023
	7173	10101101000	6747	01000011111	87(105)	4	703915
	7107	01101101011	7047	01001100011	91(121)	2	698315
	7647	10010010010	7137	11100011101	87(111)	8	690739
	7627	01001010010	7237	11100101001	95(115)	2	702371
	7467	10100011001	7317	11111110011	87(117)	4	702171
$N = 12, p = 4095$	14501	101111000001	10123	010011000000	131(167)	4	2825203
	11301	101011011011	10151	101111000000	141(173)	2	2701423
	15701	110111101000	10173	111001101000	129(155)	2	2749847
	13701	110010011001	10175	110011110100	125(157)	6	2790223
	11441	010011001000	10231	001010010101	141(173)	2	2780595
	10541	111110000000	10321	100111001111	137(171)	4	2774351
	15341	010110101011	10353	110110111001	133(165)	2	2727159
	16021	110100100000	10407	101001010111	133(161)	8	2759547
	17421	110110101101	10437	010010100100	127(149)	2	2828791
	14221	110000110001	10443	001000000001	163(213)	6	2684735
	15621	001111101111	10473	111011011100	125(157)	8	2765291

17121	100001110001	10517	010010110101	129(151)	4	2808503
16521	010110010100	10527	100001001000	145(185)	2	2739575
15321	111000101001	10553	111100000101	149(193)	2	2650487
12061	111001011110	10605	110100000000	145(175)	6	2735459
14661	000101100000	10663	011101111101	137(171)	4	2640623
11561	000110101011	10731	000000011001	127(161)	2	2672127
17561	011101111011	10737	001010101100	135(171)	2	2697331
13011	100100111100	11015	000010110011	127(169)	6	2758295
16611	110110110010	11067	111100100110	139(175)	2	2674715
13611	110010110110	11075	101100101000	147(173)	2	2837723
16311	010000010101	11147	000101100101	133(159)	2	2702211
14711	100010001101	11163	100101010110	151(197)	2	2731047
17711	111100010011	11177	011111101101	131(159)	2	2655899
11651	100011111010	11271	000111111011	133(161)	8	2785779
15151	111010010000	11313	010111000010	121(143)	4	2776399
17031	011001011101	11417	010110101110	141(177)	2	2819991
13413	101000110111	11435	000010110100	137(173)	4	2753175
11631	011100111000	11471	100100110100	133(163)	4	2794039
17631	110101000010	11477	011111010100	141(191)	2	2743307
13131	000110011000	11515	010111001001	133(161)	6	2752291
14271	101001111010	11643	111001001101	143(177)	6	2733319
16005	001101110010	12007	110111111000	157(197)	4	2792291
16605	001101001001	12067	101000011110	135(171)	2	2700603
17105	010001110011	12117	000101100011	137(189)	8	2761503
13505	100001011011	12135	000001100101	137(169)	2	2696255
16305	011010110110	12147	011011000100	127(149)	2	2712847
12705	001101110010	12165	110000110110	141(169)	2	2790467

Table 3.1 *Continued*

Sequence	Loading	Sequence	Loading	Min (max)	C	SE
16245	111000001011	12247	111111101100	139(171)	2	2799139
13245	111000111010	12255	100110110110	147(175)	2	2820047
14545	100100000110	12323	101010011000	135(165)	4	2777259
17025	100000110000	12417	000100011000	149(177)	2	2778367
13425	110101000011	12435	101001010011	137(155)	2	2609955
13125	010101011110	12515	111111101100	123(151)	4	2786951
14465	111011101100	12623	001011101000	137(171)	2	2728171
16565	011110100010	12727	011101110001	139(169)	4	2820263
13565	011101101010	12735	011000001000	131(157)	8	2781659
15365	000000010000	12753	010011000010	133(171)	2	2720419
16115	100001100101	13107	101010110101	141(181)	2	2855907
13655	000000011001	13275	100011111001	135(165)	6	2728511
14135	011101000000	13503	000110110100	139(163)	2	2826207
14675	001010001001	13663	110011100100	129(155)	4	2759251
17675	000110001101	13667	110101011001	133(169)	8	2779483
16503	111000111011	14127	011101010100	127(149)	14	2793143
16443	101100011000	14227	001101001100	141(173)	2	2703699
17343	101110100100	14357	010111001100	143(179)	2	2786731
15423	101010100111	14433	111101010010	129(153)	2	2764799
15723	100011011111	14553	011001101001	141(177)	4	2780167
15063	111000011011	14613	010001100100	129(163)	6	2792235
17163	110001101111	14717	011000101111	129(155)	6	2873167
16363	000000111011	14747	111010101100	135(165)	2	2772067
15413	011001100100	15033	101100011011	135(161)	10	2752263
15213	101100110110	15033	010000011011	131(177)	6	2813935

17433	110100100101	15437	011100001001	131(167)	2	2768319
16533	000010000010	15527	111110001100	133(159)	6	2916739
16273	101110001001	15647	100011011110	141(163)	2	2710175
17673	000000010010	15677	001010110100	133(157)	2	2814083
16407	100110011000	16027	000111001000	139(167)	2	2756595
16207	000011010110	16047	010000010111	133(165)	8	2803023
17447	000000101110	16237	111000100011	125(153)	2	2099271
17147	001100000011	16317	000111010001	153(185)	8	2724651
17217	101101010011	17057	000101011000	131(165)	2	2712235

(Reprinted from 'Numerical evaluation of the nonperiod autocorrelation parameter for optimal phases of maximal length sequences', *IEE Proceedings F*, vol. 127, no. 3, pp. 230–237, June 1980.)

with F_{opt} gives the result shown in Fig. 3.6. The merit factors for the *m*-sequences are down by a factor of between 1.8 and 3.2, and get worse as the sequence period is increased.

For a random binary sequence the average F value becomes $p^2/(p-1)p \cong 1$ for p values of interest. Thus optimal *m*-sequences perform considerably better with respect to the merit factor than true random sequences.

Table 3.2 *Least energy sequences maximized for auto-optimality with periods from 31 to 255*

	Sequence	Loading	Sequence	Loading	*SE*	Min	*C*
(a)	$N = 5, p = 31$						
	45	10011	51	10100	103	9	2
	67	11000	73	01010	115	9	2
	75	11110	57	10010	91	7	2
(b)	$N = 6, p = 63$						
	103	000010	141	011111	427	11	2
	133	100010	155	101100	492	11	4
	147	110001	163	101011	351	13	2
(c)	$N = 7, p = 127$						
	211	1100100	221	0100111	1915	21	2
	217	0000101	361	1111111	2015	15	12
	235	0110000	271	0010001	2107	21	4
	247	0010111	345	0110001	2255	17	8
	277	1001101	375	0000111	2167	21	2
	357	1011110	367	0000010	2199	21	4
	323	1111011	313	0001110	2055	21	2
	203	0110111	301	1011011	2043	21	2
	325	0101011	253	0100101	2191	27	2
(d)	$N = 8, p = 255$						
	455	11011110	551	00000011	9463	29	2
	453	11111111	651	01001000	9099	27	6
	435	11000010	561	11111100	9059	27	4
	537	10110100	765	00111100	8833	25	8
	545	00110011	515	10110110	9383	29	2
	543	01000110	615	10101001	9215	33	2
	607	11100001	703	10111011	9223	29	2
	717	01110011	747	00111100	8899	31	2

(PURSLEY, M. B. and ROEFS, H. F. A., 'Numerical evaluation of correlation parameters for optimal phases of binary shift-register sequences', *IEEE Trans.*, 1979, **COM-27** (10), pp. 1597–1604.

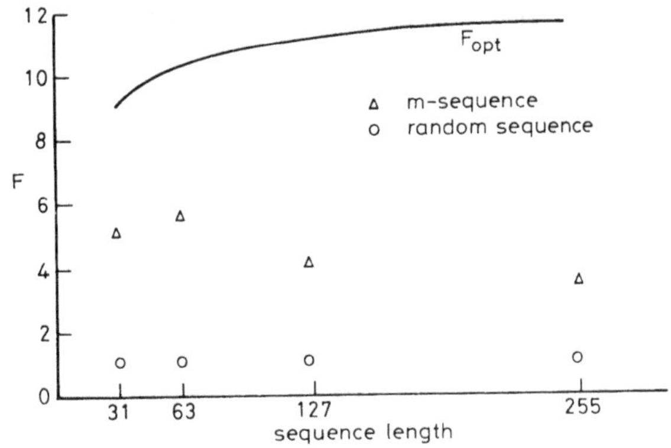

Fig. 3.6 *Merit factor of binary sequences*

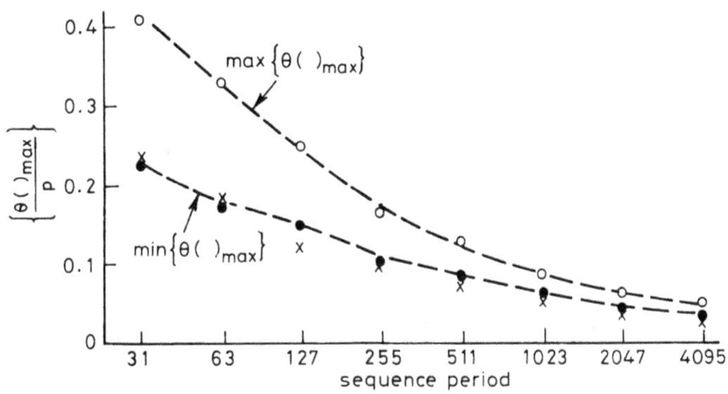

Fig. 3.7 *Odd auto-correlation parameters for maximal length sequences of periods 31 to 4095. The curves for max $\theta(\ )_{max}$ is obtained from the maximum among all cyclically distinct m-sequences of each period. The curve showing min $\theta(\ )_{max}$ is the optimization of the sequence used to plot max $\theta(\ )_{max}$ and thus show the 'best' auto-optimal sequence among all cyclically distinct m-sequence of each period*
(Reprinted from 'Numerical evaluation of the nonperiod autocorrelation parameter for optimal phases of maximal length sequences', *IEE Proceedings F*, vol. 127, no. 3, pp. 230–237, June 1980.)

The computational effort in deriving the results presented in Table 3.1 is considerable even when special algorithmic procedures are applied [14]. The natural question then to ask is at what point the results cease to be practical and become largely academic. Fig. 3.7 shows a plot for max $|\hat{\theta}^k(l)_{max}|/p$ among all cyclically distinct m-sequences for each different sequence period. The curve clearly shows the steady decrease in the ratio as p is increased, and in addition how the difference between the 'best' and the 'worst' phase for each phase becomes smaller as p is increased. In [9] it is shown that the ratios will tend to zero as p tends to infinity, and thus an optimization of $\hat{\theta}^k(l)$ might be expected to be of less importance in systems using longer sequence periods than considered here.

3.6.3 *Periodic crosscorrelation of maximal length sequences*

In systems using mark–space signalling, or in code division multiple access applications, it has been shown that the periodic and nonperiodic crosscorrelation functions are of major importance. The periodic crosscorrelation will be considered first. Since no analytical expression is known which can be used to calculate the crosscorrelation function between two particular m-sequences, the values must be computed by performing multiplication bit by bit and adding the result. This is a somewhat impractical solution, and it would at least be attractive if it could be shown statistically that the fluctuation in the crosscorrelation value for a pair of sequences was small for all relative phase shifts so that the mean could be used representatively. However, in Appendix B it is shown that the mean of the periodic crosscorrelation for all shifts is given by $1/p$ and the variance for all shifts is approximately given by $1 + (1/p)$.

For example, for m-sequence pairs of period $p = 1023$ the maximum crosscorrelation value may vary from 65 to 383 from pair to pair. However, if the sequence pair with the lower crosscorrelation bounded by 65 is chosen there is still a fluctuation from relative phase shift to phase shift of $1 \leqslant |\hat{\theta}^{kr}| \leqslant 65$. Thus the mean is not very representative for the periodic crosscorrelation from relative phase shift to phase shift. In order to indicate what can be expected for the crosscorrelation a few lower bounds can be established. From the result of Gold [4] developed in Appendix A,

$$\sum_{l=0}^{p-1} [\theta^{kr}(l)]^2 = \sum_{l=0}^{p-1} \theta^k(l)\theta^r(l) \tag{3.28}$$

The left-hand side of eqn 3.28 is upper bounded by $p \max(\theta^{kr})$. Furthermore, the right-hand side cannot be less than $p^2 - (p - 1)\max(\theta^k)\max(\theta^r)$. Therefore

$$p \max(\theta^{kr}) > p^2 - (p - 1) \max(\theta^k) \max(\theta^r)$$

Since m-sequences are considered, the maximum value of the autocorrelation outside the main peak is -1, so that

$$\max(\theta^{kr}) > (p - 1)^{1/2} \tag{3.29}$$

A much tighter bound due to Sidel'nikov [15] is valid for a set of K sequences where $K > p$. Then the maximum crosscorrelation between any sequence pair taken from this set is lower bounded by

$$\max(\theta^{kr}) > (2p - 2)^{1/2} \tag{3.30}$$

A designer's approach to the avoidance of large mutual crosscorrelation has been to examine the factors of the period of the sequences [16]. If it was found that the sequence period had small factors, the chance of large values of crosscorrelation between some pairs was considered high. It is however a rather drastic approach to use this guide and totally exclude all sequences with periods having small factors. A closer look at sequences which have small factors does however offer a general sieve to remove only those particular sequence pairs which are likely to exhibit large crosscorrelation. It can be shown that it is the combination of sequence period factors and the decimation property of *m*-sequences which makes possible large crosscorrelation values. Let q denote a positive integer, and consider the sequence $\boldsymbol{a}^r$ formed by taking every qth element of the *m*-sequence $\boldsymbol{a}^k$. The sequence $\boldsymbol{a}^r$ is said to be a decimation by q of $\boldsymbol{a}^k$. If $(p, q) = 1$, the decimation is called proper and the sequence $\boldsymbol{a}^r$ of period p is another *m*-sequence. In this way it is possible to construct all *m*-sequences of a particular period by proper decimations of one *m*-sequence.

The crosscorrelation between two sequences $\boldsymbol{a}^k$ and $\boldsymbol{a}^r$ can now be defined as

$$\theta^{kr}(i) = \sum_{l=0}^{p-1} a_l^k a_{l+i}^r = \sum_{l=0}^{p-1} a_l^k a_{q(l+i)}^k \tag{3.31}$$

This means that the crosscorrelation between $\boldsymbol{a}^k$ and $\boldsymbol{a}^r$ for any i can be obtained through the process of multiplying the digits a_l and a_{ql}, $l = 1, 2, \ldots, p$, and summing over the period where $(p, q) = 1$.

To obtain the mutual crosscorrelation for the different relative phases i between the *m*-sequence pair, it is necessary to carry out the decimation on all the cyclic permutations. However, the decimation by any particular $q = q_1$ on the cyclic permutations $T^s\boldsymbol{a}^k$, $s = 0, 1, \ldots, p - 1$ will in general not lead to all possible relative phase positions between $\boldsymbol{a}^k$ and $\boldsymbol{a}^r$. The complete result is obtained through further decimations of the form $q_2 = q_1 2^h \bmod p$ for positive integers h. The number c of times T^s can be used on the original sequence $\boldsymbol{a}^k$, after which decimation by q_1 results in a new phase for $\boldsymbol{a}^r$, is given by

$$c\,q \bmod (y\,p) = c \tag{3.32}$$

where y is any positive integer. Eqn 3.32 can also be written as

$$\frac{c}{y} = \frac{p}{q - 1} \tag{3.33}$$

The number of times T^s can be used on the original sequence $\boldsymbol{a}^k$, after which decimation by q_1 does not result in a new phase for $\boldsymbol{a}^r$, is a direct measure of how often the sequence digits $a_l \equiv a_{q_1 l}$. The number of times $a_l \equiv a_{q_1 l}$ equals $p/c = |D|$, where $|\ |$ denotes the number of elements in the set D, so that

$$\sum_{l \in D} a_l^k a_{q_1 l}^k = +|D|$$

because $a_l^k a_{q_1 l}^k$, $l \in D$ will give $(-1)(-1)$ or $(+1)(+1)$, both of which give $+1$ as the result. For cases where $|D|$ is large, the crosscorrelation will have a large positive bias, which is sufficient for one to expect the correlation to exhibit large peak values.

Example 3.2
For $p = 63 = 3 \times 3 \times 7$ and a proper decimation of $q_1 = 22$, eqn 3.33 gives

$$\frac{c}{y} = \frac{63}{21} = \frac{3}{1}$$

$|D| = 63/3 = 21$, and one would expect large crosscorrelation values between the m-sequence pair.

For the same p but with another proper decimation (giving another m-sequence) of $q_1 = 7$, eqn 3.33 gives

$$\frac{c}{y} = \frac{63}{6} = \frac{21}{2}$$

$|D| = 63/21 = 3$, and one would not expect particularly large crosscorrelation values. The calculation of $|D|$ not only gives a good sieve for identifying sequence pairs with high mutual crosscorrelation values, but also can be used to identify the number of high correlation peaks. It can be shown [17] that $\theta^{kr}(i) = \theta^{kr}(j)$ whenever i and j belong to the same cyclotomic coset. Now, for each member formed under multiplication of the subgroup $(1, 2, 4, 8, \ldots)$ and $|D|$, the results taken mod p forms a coset. The number of elements in the coset equals the number of times the crosscorrelation may rise to the particular peak value. When $|D|$ is 21 the two elements in the coset are 21 and 42 and two high correlation peaks can be expected. To verify this, two code generations given by $f_A(x) = x^6 + x^5 + 1$ and $f_B(x) = x^6 + x^3 + x^2 + x + 1$ respectively were considered. f_B produces a sequence identical to the sequence obtained by decimating the sequence generated by $f_A(x)$ by 22. The crosscorrelation results are shown in Fig. 3.8.

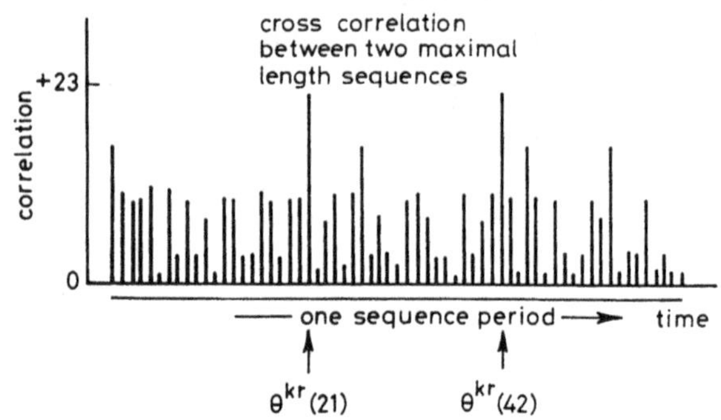

Fig. 3.8 *Cross-correlation between two maximal length sequences giving a few high peak values*

Gold [18] describes in his paper an analytical technique which tells how to select m-sequences with a specified upper bound on the crosscorrelation function. As will be shown these upper bounds are very close to the bounds in eqn 3.30 and will therefore be near to optimum. Gold's results are based on the following theorem:

If θ is any primitive element of $GF(2^m)$ and f_1 is the minimal polynomial of θ, then let f_t be the minimal polynomial of q^t, where

$$t = 2^{(m+e)/2} + 1 \qquad e = 1, n \text{ odd} \qquad e = 2, n \text{ even} \tag{3.34}$$

If now k is the m-sequence generated by the polynomial f_1 and r is the m-sequence generated by the polynomial f_t, the crosscorrelation function between sequences k and r will be three valued and bounded by t. Thus

$$k \in V(f_1) \text{ and } r \in V(f_t) \rightarrow |\theta^{kr}| \leqslant t \tag{3.35}$$

The two m-sequences k and r are called a *preferred pair* of m-sequences.

Example 3.3

Let α be a root of f_1. Thus from the table of irreducible polynomials in [10] f_1 is given in octal notation as 2011, which means $f_1 = x^{10} + x^3 + 1$. The value of t is calculated to give $2^{(10+2)/2} + 1 = 65$. It is now required to find the minimal polynomial of α^{65}. The table in [10] has no entry of degree 10 for which α^{65} is a root. However, 17 belongs to the same coset as 65 and the minimal polynomial for 17 is listed as 3515 in octal notation, giving $f_t = x^{10} + x^9 + x^8 + x^6 + x^3 + x^2 + 1$. The preferred pair of m-sequences generated by f_1 and f_t will have a crosscorrelation bounded by 65, and the three values of the crosscorrelation can be obtained from the general result

$$\theta^{kr} = -1 \quad \text{or} \quad -1 \pm 2^{(m+e)/2} \qquad \begin{array}{l} e = 1,\ m \text{ odd} \\ e = 2,\ m \text{ even} \end{array} \tag{3.36}$$

Gold's method for constructing a preferred pair of m-sequences gives no solutions when m is a multiple of 4. However, a relaxation of the requirements for the crosscorrelation to be three valued enables the construction of sequence pairs which also achieve similar bounds on the crosscorrelation function as preferred pairs when m is a multiple of 4. Table 3.3 shows the crosscorrelation between all m-sequence pairs with periods from 63 to 1023 divided into two categories, those who satisfy the upper bound from Gold's theorem and those who do not.

No restriction on the correlation magnitudes exists except that the largest value should satisfy the bound from Gold's theorem.

Notice that some of the sequence pairs that do satisfy Gold's bound are reciprocal sequences. In fact it can be shown in general for reciprocal m-sequences that their crosscorrelation functions are bounded by $\theta^{kr} < 2^{(m+2)/2}$ [19]. The table also shows that there exist several sequence pairs satisfying the correlation bound given by Gold's theorem also for $m = 8$ where no preferred pairs can be constructed from the theorem.

3.6.4 *Generation of Gold codes and their correlation properties*

A closer look at the matrixes in Table 3.3 clearly shows that even though many sequence pairs satisfying Gold's theorem exist, it is not possible to construct large sets of m-sequences satisfying the bound and thus it is difficult to design a multiple access system of reasonable size.

The significance of Gold's theorem is, however, that a preferred pair of m-sequences can provide large families of nonmaximal sequences with a bound

Table 3.3 *Crosscorrelation for m-sequences with periods from 63 to 1023. The numbering of the sequences refers to the list of feedback taps given in the right-hand column*

Crosscorrelation matrix for 6 *m-sequences, N* = 6
0 = correlation greater than 17
1 = correlation less than 17

	1 I 10 I	Feedback taps
1	11100	6,1
2	0011	6,4,3,1
3	011	6,5
4	11	6,5,2,1
5	0	6,5,3,2
6		6,5,4,1

Crosscorrelation matrix for 18 *m-sequences, N* = 7
0 = correlation greater than 17
1 = correlation less than 17

	1 I 10 I 20 I	Feedback taps
1	11101111001010010	7,1
2	1011101101101000	7,3
3	011111001101000	7,3,2,1
4	00110110110111	7,4
5	1001110101101	7,4,3,2
6	001111101000	7,5,2,1
7	11001010110	7,5,3,1
8	1001010111	7,5,4,3
9	001001010	7,5,4,3,2,1
10	10111101	7,6
11	0111111	7,6,3,1
12	011010	7,6,4,1
13	01101	7,6,4,2
14	0111	7,6,5,2
15	001	7,6,5,3,2,1
16	11	7,6,5,4
17	1	7,6,5,4,2,1
18		7,6,5,4,3,2

on their crosscorrelation. Since the sequences are nonmaxial their autocorrelation function will not be two valued; however, the out-of-phase value of the autocorrelation function will satisfy the bound from Gold's theorem. Thus by relaxing somewhat the condition on the autocorrelation function a large family of sequences with bounded crosscorrelation can be found. The sequences generated from a preferred pair of m-sequences are usually called Gold codes. The Gold code generator is simply constructed either by using the feedback shift register corresponding to the product polynomial $f_1 f_t$, or by modulo 2 adding the outputs from the two feedback shift registers corresponding to f_1 and f_t. In both cases a family of $2^m + 1$ sequences can be generated with a period of $2^m - 1$ and with mutual auto- and crosscorrelation bounded by t.

Table 3.3 *Continued.*

Crosscorrelation matrix for 16 *m-sequences, N* = 8
0 = correlation greater than 33
1 = correlation less than 33

	1 I 10 I 20 I	Feedback taps
1	010000010000001	8,4,3,2
2	00100000110000	8,5,3,1
3	0001100000000	8,5,3,2
4	001100000010	8,6,3,2
5	00010000001	8,6,4,3,2,1
6	0001110000	8,6,5,1
7	000000100	8,6,5,2
8	10000000	8,6,5,3
9	0000000	8,6,5,4
10	001100	8,7,2,1
11	01000	8,7,3,2
12	0001	8,7,5,3
13	010	8,7,6,1
14	10	8,7,6,3,2,1
15	0	8,7,6,4,2,1
16		8,7,6,5,4,2

Example 3.4
Consider the polynomials generated in Example 3.3. The Gold code generator can now be made either from a feedback shift register corresponding to the polynomial, $f_1 f_t = x^{20} + x^{19} + x^{18} + x^{16} + x^{11} + x^8 + x^5 + x^2 + 1$, or by modulo 2 adding the outputs from the feedback shift registers corresponding to $f_1 = x^{10} + x^3 + 1$ and $f_t = x^{10} + x^9 + x^8 + x^6 + x^3 + x^2 + 1$. Both

Table 3.3 *Pt. 1.*

Crosscorrelation matrix for 48 *m-sequences, N* = 9
0 = correlation greater than 33
1 = correlation less than 33

	1 I 10 I 20 I 30 I 40 I 50 I	Feedback taps
1	100010001000011001010000001001000000111000000000	9,4
2	00100110011000000001100000000000001000010100010	9,4,3,1
3	0100100011000100000000000000011100000110110000	9,5
4	011000100010010010001110000000100010000001000	9,5,3,2
5	00001000000000000000110011000101110000000001	9,5,4,1
6	0010001100000001110000000010000000001110000	9,6,4,3
7	010001100000001100000100100010000001110000	9,6,4,3,2,1
8	00100011011110000101001000000000000001000	9,6,5,3
9	0100011001100001100010000000010000000000	9,6,5,3,2,1
10	010100100000010000100001010000010100100	9,6,5,4,2,1
11	00000100000101000010011000000001000110	9,6,5,4,3,2
12	0011011010000001001000100000000000001	9,7,2,1
13	001010100000000001000000110000001001	9,7,4,2
14	01010100000110010000100010000001000	9,7,5,1
15	0000001000000000110010000010010010	9,7,5,2
16	000001000000000001010000010100010	9,7,5,3,2,1
17	10110001011011000100000000000000	9,7,5,4,2,1
18	0000100000000100000000001110010	9,7,5,4,3,2
19	001000000000010001000111010010	9,7,6,3,2,1
20	01110000100001000000000000110	9,7,6,4
21	0010000010001001000110010100	9,7,6,4,3,1

22	00100001000000101000000000	9,7,6,5,4,2
23	0011001100010001000001000	9,7,6,5,4,3
24	110101100000011000000000	9,8,4,1
25	00010000000000101010110	9,8,4,2
26	0010001101000101000000	9,8,4,3,2,1
27	000001100000111000100	9,8,5,1
28	10011001000111000100	9,8,5,4
29	1100000101000000000	9,8,5,4,3,1
30	000000000000100110	9,8,6,3,2,1
31	00101000010000100	9,8,6,4,3,1
32	0000110000001001	9,8,6,5
33	010111000000001	9,8,6,5,3,1
34	10111000000001	9,8,6,5,3,2
35	0000101000110	9,8,6,5,4,1
36	110000000000	9,8,6,5,4,3,2,1
37	00100110000	9,8,7,2
38	0100110000	9,8,7,3,2,1
39	101000000	9,8,7,5,4,2
40	00000001	9,8,7,5,4,3
41	0001001	9,8,7,6,2,1
42	001000	9,8,7,6,3,1
43	01001	9,8,7,6,3,2
44	1001	9,8,7,6,4,2
45	110	9,8,7,6,4,3
46	01	9,8,7,6,5,1
47	1	9,8,7,6,5,3
48		9,8,7,6,5,4,3,1

Table 3.3 *Pt. 2.*

Crosscorrelation matrix for 60 *m-sequences, N* = 10
0 = correlation greater than 65
1 = correlation less than 65

	1 I 10 I 20 I 30 I 40 I 50 I 60 I	Feedback taps
1	000011100100101000000000001000000000100000000000011000000100	10,3
2	00010000001000000000100000100010000100011110001000000000000	10,4,3,1
3	0000110000000001000000000000010000000010100101010001 10000	10,5,2,1
4	0000100000010110000110000000001000100000000010100000 1000	10,5,3,2
5	000000000101001001100001010001000000000100000001 1000000	10,6,5,2
6	0100000010000000001100001000000000000100000101000 01000	10,6,5,3,2,1
7	000010000010101010000000100000010001000010000000 01000	10,7
8	00000100000000000111000000000000100100000000010000 0100	10,7,3,1
9	1000110110000000000000000000100000000000101000001000	10,7,6,2
10	100000011000110000000000000000100000000010110000 1000	10,7,6,5,4,3,2,1
11	00100000000100100000000000100011000000010000010001 00	10,7,6,4,2,1
12	0000000100000000001000001100110000000000000011100 10	10,7,6,5,2,1
13	0000000101001010000011001000100000001000000000001	10,7,6,5,4,1
14	010000010001000000000000101100000000010000000000	10,8,3,2
15	00001010010100000000000000000010100001000000 10	10,8,4,3
16	0000000001000001001100011000000000000001000100	10,8,5,1
17	0100000000100001000000000001000011100011 0000	10,8,5,4
18	10000100100000000000011000001000000010000000	10,8,5,4,3,2
19	0000000000010010000000000000000101000001100	10,8,6,1
20	000010001010000000000000000100011100000 00	10,8,6,4,2,1
21	10000110001100010010110000100000000000 00	10,8,6,5,3,1

22	01000000010000001000000101000000000010	10,8,7,2
23	0000010000100110000000010000010100000	10,8,7,3,2,1
24	000000000000000000100100010001001000	10,8,7,4,2,1
25	01000110000000000001000010000010000	10,8,7,5
26	1000000000000000000100110000000100	10,8,7,6,2,1
27	000101000000000000000000101000000000	10,8,7,6,5,2
28	00010100110000000001010001000000	10,8,7,6,5,4,2,1
29	0000110010010011010000000001000	10,8,7,6,5,4,3,1
30	101101000000110000000001110001	10,9,4,1
31	00000000000101010000100000000	10,9,4,2
32	0000000001100100000000010000	10,9,5,2
33	000000000000000101010110010	10,9,5,4,2,1
34	10100001000000000010000100	10,9,6,1
35	0110000000010000010010000	10,9,6,3,2,1
36	001111000010000000000011	10,9,6,4,3,1
37	10001000000001000000000	10,9,6,5,4,3
38	0001000000000010000010	10,9,6,5,4,3,2,1
39	000100110000000001000	10,9,7,3
40	00000000000011000101	10,9,7,5,4,2
41	0000000000001001101	10,9,7,6
42	000001000000000000	10,9,7,6,4,1
43	00101000000000100	10,9,7,6,4,3,2,1
44	0000101000110010	10,9,7,6,5,4,3,2
45	110100110000001	10,9,8,4,2,1
46	01000000001001	10,9,8,4,3,2
47	0000010000000	10,9,8,5
48	000100100000	10,9,8,5,4,3

Continued.

Table 3.3 *Pt. 2 continued.*

49	00000000000	10,9,8,6,2,1
50	0100000100	10,9,8,6,3,2
51	000000000	10,9,8,6,4,2
52	00100000	10,9,8,6,4,3
53	0000000	10,9,8,6,5,1
54	000011	10,9,8,6,5,4,3,2
55	00011	10,9,8,7,3,2
56	0011	10,9,8,7,4,1
57	000	10,9,8,7,5,4
58	00	10,9,8,7,6,4,3,1
59	1	10,9,8,7,6,5,4,1
60		10,9,8,7,6,5,4,3

solutions are shown in Fig. 3.9. A family of 1025 sequences all having a mutual cross-correlation bounded by 65 has been constructed.

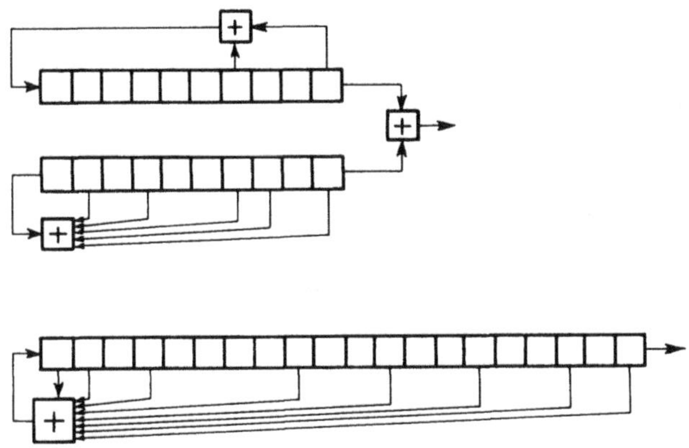

Fig. 3.9 *Mechanization of Gold code generators from example 3.4*

Since the Gold codes were not introduced during the discussion on autocorrelation parameters, a few results are presented here. The periodic autocorrelation has already been mentioned and is bounded by t from Gold's theorem. The odd autocorrelation for Gold codes has been bounded by Massey and Uhran [20]:

$$\hat{\theta}^k \leqslant 2^{n-1} + 2^{\lfloor n/2} + 2.$$

where $\lfloor n/2$ denotes the integer part of the number $n/2$. The bounds given in eqn 3.36 are not tight in general and do not take into account the sensitivity of the nonperiodic parameters to the phases of the sequences.

3.6.5 *Nonperiodic crosscorrelation properties of maximal length sequences*
To complete the discussion on spreading sequence correlation properties, the nonperiodic crosscorrelation will be considered.

Although there are no analytical expressions which give the values of $\hat{\theta}^{kr}$ for a set of m-sequences, several bounds can be applied. The result in eqn 3.25 for a set of K sequences can also be used to lower bound the odd crosscorrelation since $\theta_{\max}$ is the maximum of $\hat{\theta}^k$ and $\hat{\theta}^{kr}$. In fact eqn 3.25 is derived from an expression relating the maximum values of the odd autocorrelation and odd crosscorrelation parameters, as shown in Appendix C. Thus

$$\frac{(\hat{\theta}^{kr})^2}{p} + \frac{p-1}{p(K-1)}\left[\frac{(\hat{\theta}^k)^2}{p}\right] \geqslant 1 \qquad (3.37)$$

This seems to verify the common observation that when autocorrelation parameters are good the crosscorrelation parameters are not very good. Any of the sequences listed in Table 3.3 which satisfy the bound from Gold's theorem are upper bounded by

$$\hat{\theta}^{kr} < 2^{n-1} + 2^{n/2} + 2 \tag{3.38}$$

As for the odd autocorrelation, the odd crosscorrelation is sensitive to phase shifts. It is, however, a formidable task to find optimal phases for multiple access applications which require a large number of sequences. Pursley and Roefs [11] have calculated peak crosscorrelation parameters for some short m-sequences and Gold sequences.

If average signal/noise calculations are required, the use of eqn 3.19 simplifies the computation to involve only the nonperiodic autocorrelation.

Notice that although reciprocal sequences have periodic crosscorrelations very close to the bound from Gold's theorem and thus are attractive from a periodic viewpoint, the interference due to other users depending on $C^k(l)\,C^r(l)$ in eqn 3.19 will under certain conditions be large for such sequences. The reason is that when k and r are reciprocal sequences and are placed in phase positions so as to obtain optimal sequences with least sidelobe energy, these phase positions $(T^\alpha \boldsymbol{a}^k)$ and $(T^\beta \boldsymbol{a}^r)$ are unique and such that $\alpha + \beta = p + m$. Under this condition $C^k(l) = C^r(l)$ and the product becomes positive in the summation $\sum_{l=1}^{p-1} C^k(l)\,C^r(l)$.

If however some of the sieves used to find optimal sequences with least sidelobe energy are removed, more than one possible phase exists and it is possible to obtain $\alpha + \beta \neq p + m$. In this case reciprocal sequences will not behave differently from any other sequence pair.

3.6.6 *Multiple interference in systems using maximal length and related sequences*

Some insight can be obtained into the interference parameter, without an actual calculation of the crosscorrelation of the sequences, if the sequences are assumed to be random. By random is meant sequences of independent, identically distributed random variables for which $P_r(+1) = P_r(-1) = 1/2$. This should be a fair assumption since the m-sequences have many characteristics in common with random binary sequences.

Calculation of the first and third moment of the interference parameter S^{kr} in eqn 3.17 using random binary sequences gives [21]

$$E[S^{kr}] = \frac{K-1}{3p}$$

and

$$\text{var}\,[S^{kr}] = \frac{K-1}{6p^3}$$

This indicates a rather small value for the interference fluctuations even for modest sequence lengths. Thus in the preliminary design stage the approximation of eqn 3.19

$$SNR = \left[\frac{K-1}{3p} + \frac{N_0}{2E}\right]^{-1/2} \tag{3.39}$$

is very useful. However, to complete the analysis for a chosen sequence set requires the use of eqn 3.19.

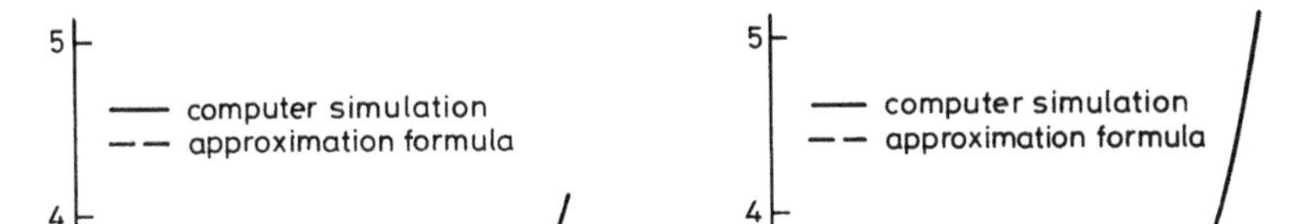

Fig. 3.10 *Performance for spread spectrum multiple access using m-sequences of period* *a* 63 *b* 127 *c* 255 *d* 511 *e* 1023

The value of *SNR* from eqn 3.19 has been simulated by computer for *m*-sequences in a phase shift where the peak nonperiodic autocorrelation is minimized (i.e. only sieve (a) in Section 3.6.2 is applied). The simulation has been carried out for $p = 63$ up to and including $p = 1023$. The result is plotted in Fig. 3.10. These graphs show degradation in *SNR* due to code interference as a function of E/N_0 for various numbers of users *K*. In addition the degradation in *SNR* is calculated using eqn 3.39. The simulation shows clearly that the approximation eqn 3.39 is very useful for most values of *p* and the number of users *K*. However, for small *p* or *K* (or both) there will be certain differences between the *SNR*s calculated from eqs 3.19 and 3.39. Results for a selected set of 60 nonmaximal codes satisfying the bounds from Gold's theorem with period 1023 is shown in fig. 3.11. The sequences have $|\theta^k| < 65$, $57 < |\hat{\theta}^k| < 71$ and $|\hat{\theta}^{kr}| < 65$.

3.6.7 *Crosscorrelation parameters between m-sequences with different periods*
Situations where sequences of different periods will occur at the same time and

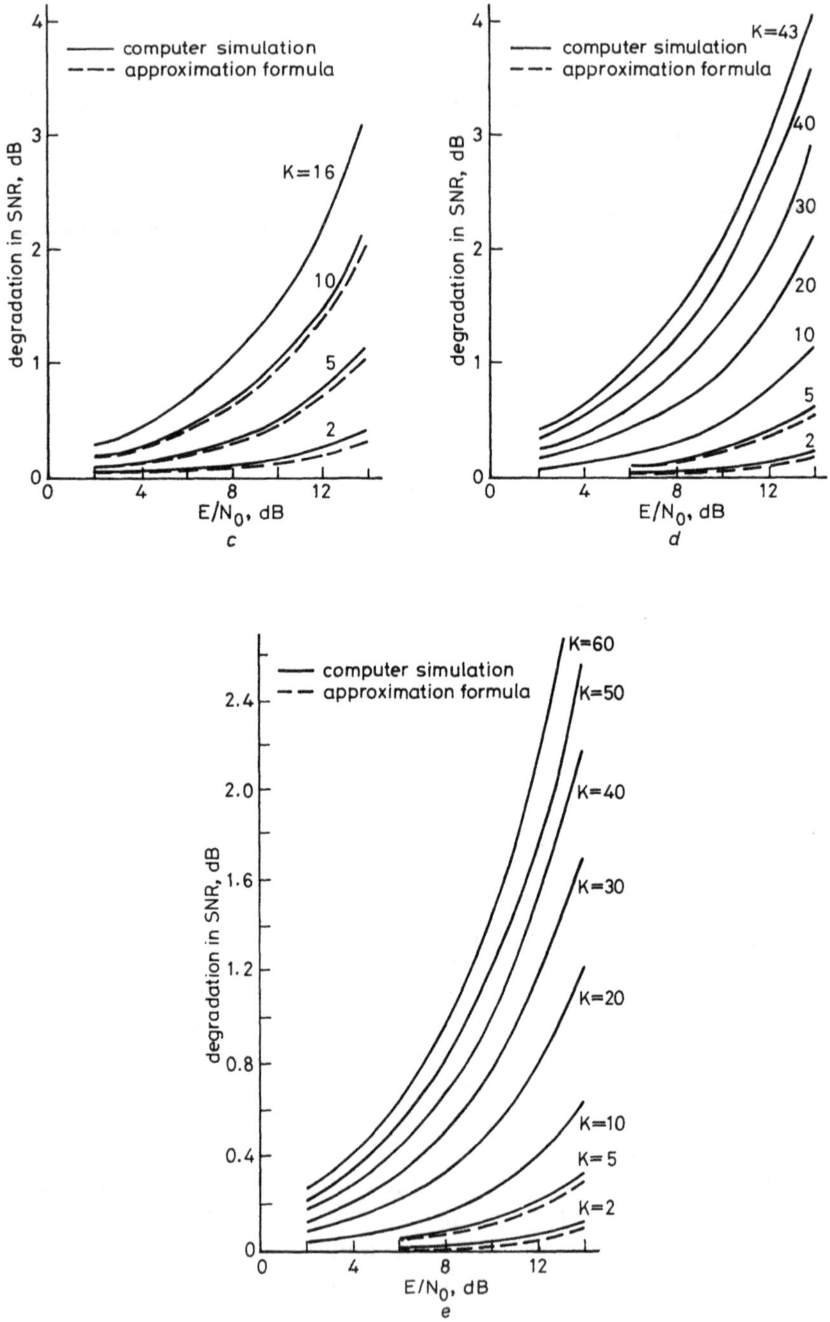

Fig. 3.10 *Continued.*

at the same carrier frequency, and therefore interfere, may exist for example in a multiple access system using different m-sequences for synchronization preambles and data spreading purposes. There is no general set of formulas or bounds which can be applied to provide an explicit analysis of this problem.

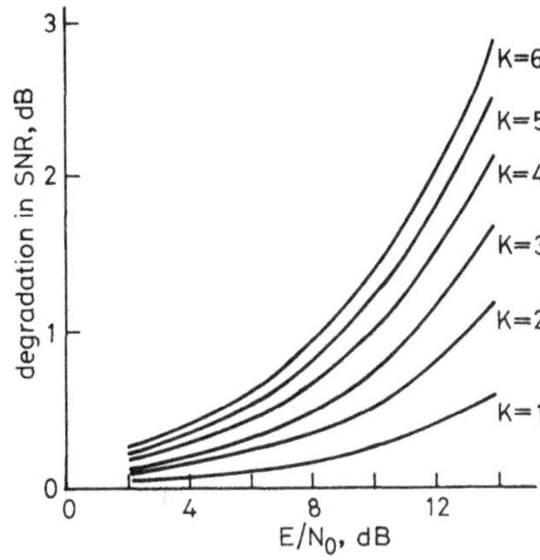

Fig. 3.11 *Performance for spread spectrum multiple access using nonmaximal sequences satisfying the bound from Gold's theorem. Period = 1023*

However, if by factorization it is shown that the periods of the different m-sequences are relatively prime, there exist methods to calculate parameters indicating the degree of crosscorrelation. A system using a synchronization preamble of $p = (2^7 - 1)$ but $p = (2^{10} - 1)$ for data spreading, and a system with a preamble of $p = (2^{10} - 1)$ but $p = (2^{19} - 1)$ for data spreading, would both use sequence lengths, being relatively prime. In this case the analysis developed for linear product sequences [23, 24] can be applied to indicate the crosscorrelation between m-sequences with relatively prime periods.

Consider two m-sequences $\boldsymbol{a}^k$ and $\boldsymbol{a}^r$, where the period of sequence k is p_1 and the period of sequence r is p_2, and $(p_1 p_2) = 1$ enables the construction of a product sequence defined as

$$\boldsymbol{a}^{k+r} = \boldsymbol{a}^{k'} + \boldsymbol{a}^{r'}$$

The + denotes modulo 2 addition, and the period of $\boldsymbol{a}^{k+r}$ is $p_3 = p_1 p_2$. Refer to Fig. 3.12.

The sequences $\boldsymbol{a}^{k'}$ and $\boldsymbol{a}^{r'}$ are derived from $\boldsymbol{a}^k$ and $\boldsymbol{a}^r$ by cascading respectively p_2 and p_1 copies of the sequences $\boldsymbol{a}^k$ and $\boldsymbol{a}^r$. In other words the sequence $\boldsymbol{a}^{k+r}$ will represent the crosscorrelation between the m-sequences k and r for all relative phase positions.

Such a sequence $\boldsymbol{a}^{k+r}$ is periodically well balanced since it can be shown that the number of +1s will be $(p_3 + 1)/2$ and the number of −1s will be $(p_3 - 1)/2$. However, in order to investigate the crosscorrelation between m-sequences of different periods $(p_1 p_2) = 1$, it is necessary to derive the non-periodic balance properties of the product sequence $\boldsymbol{a}^{k+r}$ using weight distributions. The first moment of the product sequence is given by

$$S^1 = \frac{M}{p^3} \tag{3.40}$$

where M is equal to p_1 or p_2.

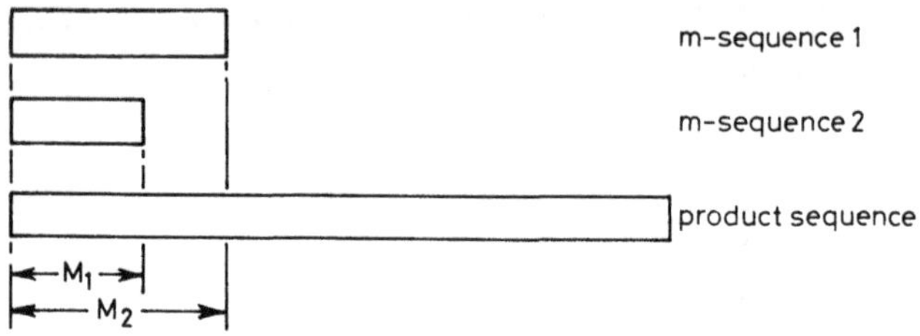

Fig. 3.12 *Construction of linear product sequences*

The second moment is derived from the shift and add properties of the m-sequences:

$$S^2 = \frac{1}{p_3}\left\{M(M + p_3 - 1) - \frac{M}{p_1}(p_1 + 1)\left[2M - p_1\left(\frac{M}{p_1} + 1\right)\right]\right.$$
$$\left. - \frac{M}{p_2}(p_2 + 1)\left[2M - p_2\left(\frac{M}{p_2} + 1\right)\right]\right\} \tag{3.41}$$

Now for $M = p_1$,

$$S^2 = \frac{1}{p_3}\left\{p_1(p_1 + p_3 - 1) - \frac{p_1}{p_2}(p_2 + 1)\left[2p_1 - p_2\left(\frac{p_1}{p_2} + 1\right)\right]\right\} \tag{3.42}$$

and for $M = p_2$,

$$S^2 = \frac{1}{p_3}\left\{p_2(p_2 + p_3 - 1) - \frac{p_2}{p_1}(p_1 + 1)\left[2p_2 - p_1\left(\frac{p_2}{p_1} + 1\right)\right]\right\} \tag{3.43}$$

The third moment becomes

$$S^3 = \frac{1}{p_3}(M^3 - q_{3k}(p_1 + 1) - q_{3r}(p_2 + 1) - q_{3(k+r)}(p_1 + p_2 + p_3 + 1) \tag{3.44}$$

where q_{3k}, q_{3r} and $q_{3(k+r)}$ are all trinomials of degree lower than M dividing the characteristic polynomials generating $\boldsymbol{a}^k$, $\boldsymbol{a}^r$ and $\boldsymbol{a}^{k+r}$ [25].

The third central moment, which gives a measure of skewness and thus might be used to indicate the degree of crosscorrelation between the two m-sequences, is given by

$$S^{3^1} = S^3 + 3S^1S^2 + 2(S^1)^3 \tag{3.45}$$

Because trinomials divisible by the characteristic polynomial generating $\boldsymbol{a}^{k+r}$ are those divisible by the characteristic polynomials generating $\boldsymbol{a}^k$ and $\boldsymbol{a}^r$, it is expected that there will be a considerable crosscorrelation if there are any low order trinomials common to both the m-sequence $\boldsymbol{a}^k$ and $\boldsymbol{a}^r$. The work involved in finding the trinomial structure for the m-sequences is generally large and requires computer simulation assistance.

3.7 Other spreading codes

The main body of this chapter on codes for bandwidth spreading has dealt with maximal length sequences and the related Gold sequences. However, many other sequence sets giving satisfactory correlation properties have been investigated in the literature. Some of these sequences have properties which make them specially attractive for particular applications.

One set of sequences derived from m-sequences is the combination sequences. Their main feature is their ability to act as quickly acquirable codes. Given m-sequences of lengths p_i, $i = 1, 2$, where the p_i are pairwise relatively prime, a combination sequence of length

$$p = \sum_{i=1}^{s} p_i \tag{3.46}$$

may be constructed. Each chip of the combination sequence is a logical combination of one chip from each of the component sequences. The choice of Boolean function used is largely decided by the autocorrelation properties obtained.

The main advantage of such sequences is however their ability to acquire a sequence of length given by eqn 3.46 with at most

$$N = \sum_{i=1}^{s} p_i \tag{3.47}$$

correlations. This differs from m-sequences, which require $\prod_{i=1}^{s} p_i$ correlations.

Another set of sequences may also be derived from m-sequences. These sequences have been introduced under several different names, but recently [26] they have been called Kronecker sequences. Consider an m-sequence $\boldsymbol{a}^k$ with period p_1 which is reproduced for each element of the m-sequence $\boldsymbol{a}^r$ with period p_2 and with its overall phase determined by the code $\boldsymbol{a}^r$ element. This produces a convoluted code of length $p = p_r p_2$. The sequence $\boldsymbol{a}^r$ is called the outer code and $\boldsymbol{a}^k$ the inner code. The height and positions of the sidelobes in the autocorrelation function of the new code will depend on the parent codes and their periods. This gives a certain freedom in the design to position the sidelobe peaks, and this might be significant in applications where multipath is a problem. There is no restriction to m-sequences as parent codes and the scheme can be expanded to any number of codes being convoluted.

Usually matched filter processing using low cost microcomputers would not be possible for any reasonable sequence length owing to speed limitations. For a sequence length of p there would be p taps, and therefore for each output p operations consisting of a memory access and an addition/subtraction would be required. However, an implementation using convoluted codes as shown in Fig. 3.13 would require only $\sqrt{p}$ additions for the case of two convoluted codes.

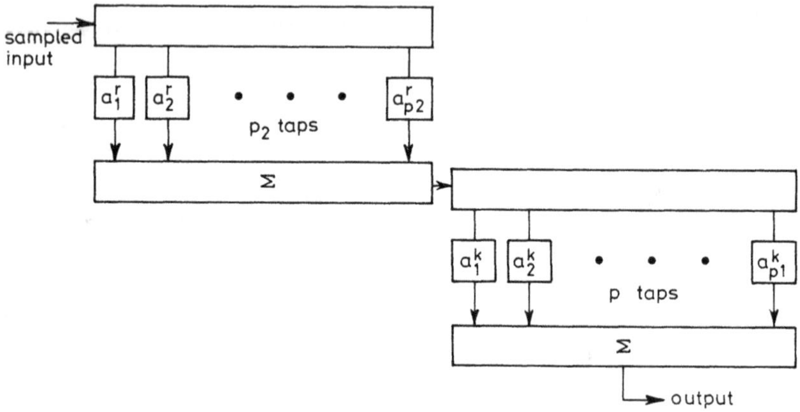

Fig. 3.13 *Matched filter algorithm for convoluted code. a_i are binary coefficients* (± 1)

The situation is generally not quite this good, however, since a certain computing overhead is required. However, a spread spectrum scheme applying convoluted sequences makes microprocessor implementation feasible (see Section 5.6). Kronecker sequences have somewhat poorer maximum correlation values than *m*-sequences but they perform about as well with average error probability. Under conditions of good *SNR* they do also allow for rapid synchronization owing to their special composition of shorter codes.

In most spread spectrum applications binary sequence are used as spreading functions, but it is possible that system performance could be improved by not restricting the sequence alphabet to binary. Recently several papers, for example [27], have focused on sequences being defined over the complex *n*th roots of unity, for some integer *n*. Several of the bounds [12] on autocorrelation and crosscorrelation which have been discussed are in fact developed in general for complex sequences. However, the construction and properties of complex sequences will not be discussed in this chapter.

3.8 Spectral occupany of *m*-sequences and Gold codes

The autocorrelation function of a signal $F(t)$ is generally defined as

$$R(\tau) = \lim \frac{1}{2T} \int_{-T}^{T} F(t)\, F(t - \tau)\, \mathrm{d}t \tag{3.48}$$

If $F(t)$ is a periodic signal like the m-sequences considered, the autocorrelation will also be periodic with the same period as $F(t)$. If now $F(t)$ is expressed as a cosine serie of the form

$$F(t) = \frac{A_0}{2} + \sum C_n \left(\frac{2\pi nt}{T} + \theta_n\right) \tag{3.49}$$

then the autocorrelation becomes a cosine series as well. The coefficients are proportional to the square of those of the original series and the phase angles θ_n are eliminated:

$$R(\tau) = \frac{A_0^2}{4} + \frac{1}{2}\sum C_n^2 \cos\frac{2\pi nt}{T} \tag{3.50}$$

It is now possible to see that all sequences whose autocorrelation is low for all values of τ except at $\tau = 0$ will tend to have a uniform harmonic spectrum. If the signal is restricted to a pattern of $+1$ and -1 combined to form a maximal length sequence, the autocorrelation may be obtained by summation of the term-by-term product of two sequences, one of them being a shifted version of the other. As defined in Section 3.6.1,

$$\theta(i) = \frac{1}{p}\sum_{l=0}^{p-1} a_l a_{l+i}$$

These sequences give uniform harmonic spectra in which, for an m-sequence of period p containing $(p - 1)/2 + 1$s,

$$C_{mn} = \frac{2}{n\pi}(p + 1)^{1/2} \sin\frac{n\pi}{p} \tag{3.51}$$

Similarly the power spectral density of the m-sequences can be obtained from the autocorrelation by

$$S(f) = \int_{-\infty}^{\infty} R(\tau)\mathrm{e}^{-j2\pi f\tau}\,\mathrm{d}\tau \tag{3.52}$$

The result is (Chapter 1):

$$S(f) = \frac{1}{p^2}\delta(f) + \frac{1+p}{p^2}\left(\frac{\sin\pi fT_c}{\pi fT_c}\right)^2 \sum_{\substack{n=-\infty \\ n\neq 0}}^{\infty} \delta\left(f - \frac{n}{pT_c}\right)$$

The resulting line spectrum has a distance between the lines of $1/pT_c$ and $\sin^2 x/x^2$ envelope. The DC component in the spectrum is due to the imbalance of $+1$s and -1s in the m-sequence.

The use of Gold codes as spreading codes may change the autocorrelation function and thus $S(f)$.

Fig. 3.14 shows the autocorrelation and power spectra for m-sequences and Gold sequences respectively. The correlation and power spectra are shown for two different Gold sequences. One of the sequences has a run distribution of

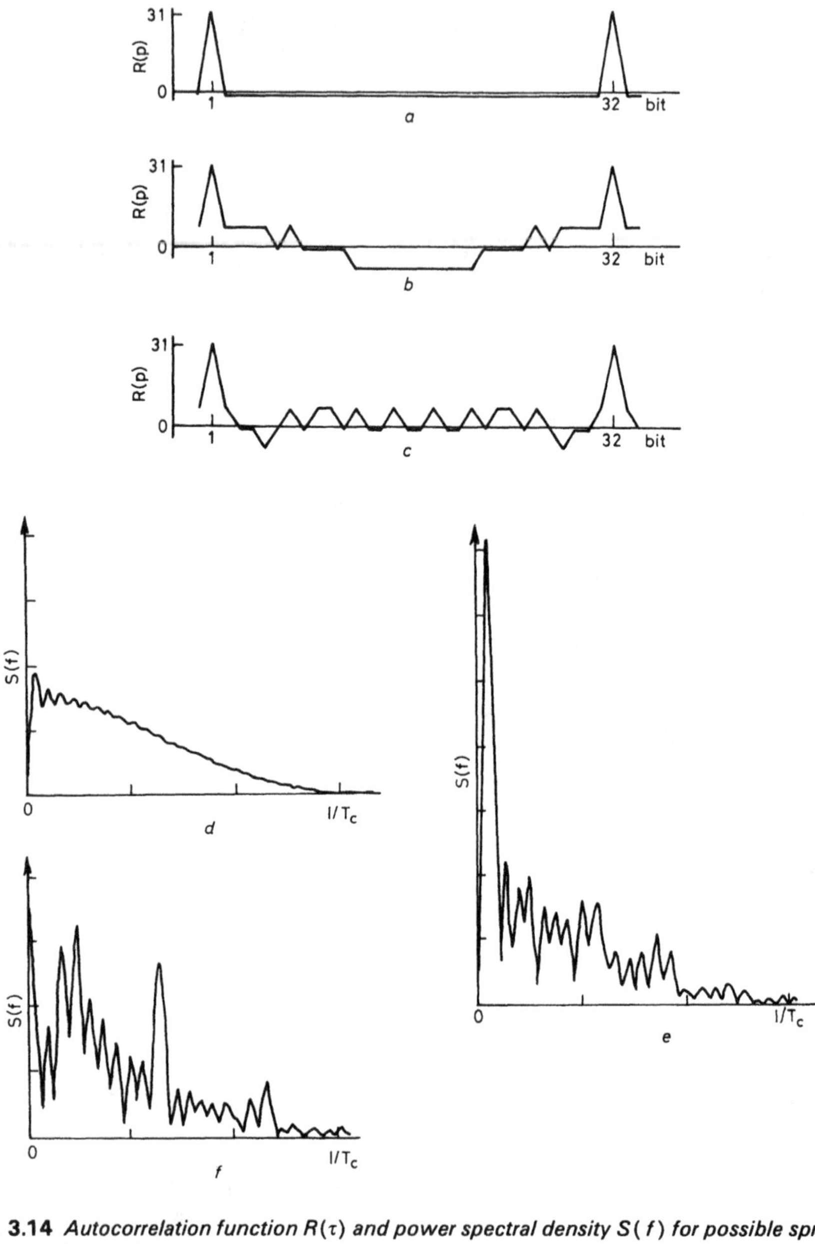

Fig. 3.14 *Autocorrelation function $R(\tau)$ and power spectral density $S(f)$ for possible spreading codes*

a $R(\tau)$ for *m*-sequence with $p = 31$
b $R(\tau)$ for Gold sequence with imbalance of -1 and $p = 31$
c $R(\tau)$ for Gold sequence with imbalance of -9 and $p = 31$
d $S(f)$ for *m*-sequence with $R(\tau)$ from (*a*)
e $S(f)$ for Gold sequence with $R(\tau)$ from (*b*)
f $S(f)$ for Gold sequence with $R(\tau)$ from (*c*)

+1s and −1s giving a total imbalance of −1, whereas the other has a total imbalance of −9. This can be seen from the DC components in the two spectra. For a system trying to achieve concealment, spectral peaks might be undesirable, since they may show up above the noise floor used to hide the spread spectrum transmissions and thus 'give away' the system.

References

1 PURSLEY, M. B.: 'Recent advances in coding for multiple access communication systems', Proceedings, International Telemetering Conference, Los Angeles, California, 1976, pp. 24–33.

2 TURIN, G. L.: 'Minimax strategies for matched-filter detection', *IEEE Trans.*, 1975, **COM-23** (11), pp. 1370–1371

3 PURSLEY, M. B.: 'Evaluating performance of codes for spread spectrum multiple access communications', 12th Annual Allerton Conference on Circuit and System Theory, Monticello III, October 1974.

4 GOLD, R.: 'Maximal recursive sequences with 3-valued recursive cross-correlation functions', *IEEE Trans.*, 1968, **IT-14** (1), pp.154–156

5 YAO, K.: 'Error probability of asynchronous spread spectrum multiple access communication systems', *IEEE Trans.*, 1977, **COM-25** (8), pp. 803–809

6 HJELMSTAD, J., and SKAUG, R.: 'Fast synchronization modem for spread spectrum communication system using burst-format message signalling', *Proc. IEE,* Part F, 1981, **128** (6), pp. 370–378

7 COLOMB, S. W. (ed.): 'Digital communications with space applications' Prentice-Hall, (Englewood Cliffs, NJ, 1964)

8 PURSLEY, M. B.: 'Performance evaluation for phase-coded spread spectrum multiple access communication. Part II: Code sequence analysis', *IEEE Trans.*, 1977, **COM-25** (8), pp. 800–803

9 SKAUG, R.: 'Numerical evaluation of the nonperiodic autocorrelation parameter for optimal phases of maximal length sequences', *Proc. IEE,* Part F, 1980, **127** (3), pp. 230–237

10 PETERSON, W. W.: 'Error-correcting codes' (Wiley & MIT Press, 1961)

11 PURSLEY, M. B., and ROEFS, H. F. A.: 'Numerical evaluation of correlation parameters for optimal phases of binary shift-register sequences', *IEEE Trans.,* 1979, **COM-27** (10), pp. 1597–1604

12 SARWATE, D. V.: 'Bounds on crosscorrelation and autocorrelation of sequences', *IEEE Trans.*, 1979, **IT-25** (6), pp. 720–724

13 GOLAY, M. J. E.: 'The merit factor of long low autocorrelation binary sequences', *IEEE Trans.*, 1982, **IT-28** (3), pp. 543–549

14 SKAUG, R.: 'Algorithm for calculating the nonperiodic autocorrelation of maximal length sequences', *Proc. IEE,* Part ECS, 1979, **3** (5), pp. 213–218

15 SIDEL'NIKOV, V. M.: 'On mutual correlation of sequences', *Soviet Math Dokl,* 1971, **12** (1), pp. 197–201

16 HARRIS, R. L.: 'Introduction to spread spectrum techniques', Agard Lecture Series No. 58, 1973.

17 GOLOMB, S. W.: 'Theory of transformation groups of polynomials over *GF*(2) with application to linear shift register sequences', *inf. Sciences,* 1968, **1** (1), pp. 87–109

18 GOLD, R.: 'Optimal binary sequences for spread spectrum multiplexing', *IEEE Trans.*, 1967, **IT-13** (5), pp. 619–621

19 McELIECE, R. J., and DOWLING, T. A.: 'Cross-correlation of reverse maximal-length shift register sequences', JPL Space Programs Summary, 1968, 37–53, Vol. 3, pp. 192–193.

20 MASSEY, J. L., and UHRAN, J. J.: 'Sub baud coding', Proceedings 13th Annual Allerton Conference on Circuit and System Theory, Monticello III, 1975.

21 ROEFS, H. F. A., and PURSLEY, M. B.: 'Correlation parameters of random binary sequences', *Electron. Lett.*, 1977, **13** (16), pp. 488–489
22 PURSLEY, M. B., and SARWATE, D. V.: 'Evaluation of correlation parameters for periodic sequences', *IEEE Trans.*, 1977, **IT-23** (4), pp. 508–513
23 TOMLINSON, G. H., and GALVIN, P.: 'Properties of the subsequences of a class of linear product sequences', *Proc. IEE*, 1975, **122**, pp. 1095–1096
24 MARITSAS, D. G.: 'On the statistical properties of a class of linear product feedback shift-register sequences', *IEEE Trans.*, 1973, **C-25**, pp. 961–962.
25 COLOMB, S. W.: 'Shift register sequences' (Holden-Day, 1967)
26 STARK, W. E., and SARWATE, D. V.: 'Kronecker sequences for spread-spectrum communication', *Proc. IEE*, Part F, 1981, **128** (2), pp. 104–109
27 ALLTOP, W. D.: 'Complex sequences with low periodic correlations', *IEEE Trans.*, 1980, **IT-26** (3), pp. 350–354
28 TORRIERI, D. J.: 'Principles of military communication systems' (Artech, 1981)

Chapter 4

Implementation

4.1 Functional requirements of communication system

The greatest challenge for the communication engineer in integrating a communication system or network is to match hardware technology and signal processing algorithms. Each step in the evolution of communication systems has usually been the result of a breakthrough in the development of a new technology or device. By contrast, changes in communication architecture and principles have within time frames of minor technological breakthroughs been those of refinement and standardization. Hence the development of communications, as in other fields of applied science, has been brought about primarily by progress in technology.

There are, however, indications that this process can change. A stage may be reached where a great variety of technologies and techniques is available to the imaginative engineer. A combination of several technologies can support systems with a performance superior to that of systems relying only on a single technology or a limited number of technologies.

There is also the issue of the increasing demand for system logistics and processor software caused by the large number of operations which have become cost effective. The requirements for effective structure of algorithms for signal processing, and for algorithms matched to the characteristics of the signal processor, put the system engineer in a position where substantial improvement in system capability can be obtained with present technology.

Then there is the question of the price of available technology. Auxiliary specifications such as high reliability and low power consumption will have to be considered; so too will the introduction of redundancy to increase reliability.

A communication system may be considered to consist of a hierarchy of computing levels, as illustrated in Fig. 4.1. The innermost level is the basic information modulation and demodulation (number crunching) level. At this level, signal processing consists of multiplications, correlations, filtering and all the adjoining manipulations. This stage is the one that regains signal/noise ratio and deciphers the data flow by applying the codes used when transmitting the spread spectrum signal.

The next level is the signal flow controlling level. Control of data shuffling operations, code synchronization and control of analogue processors are needed to support the basic signal processing.

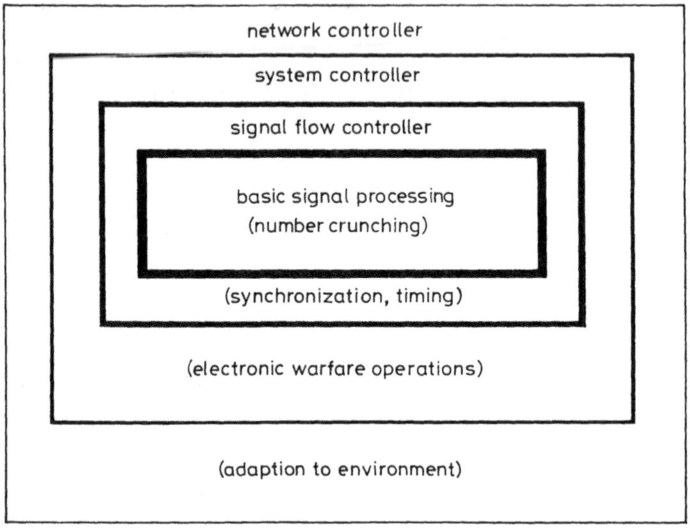

Fig. 4.1 *Hierarchy of processing levels in a communication system*

The system controller is the highest level of a communication terminal. Adaptation to the environment, ECCM and the full array of electronic warfare operations belong to this level. It is of crucial importance for any communication terminal to be able to combat signals trying to deceive the terminal. The game of electronic warfare is extremely dynamic, and the lifetime of a communication system is determined by its ability to resist jamming and intercept techniques developed after implementation of the communication system. These techniques may or may not be developed with knowledge of the principles and structure of the system under attack. It is therefore of vital importance that the communication system is sufficiently rugged to maintain an adequate resistance to jamming throughout its lifetime.

In a communication network there will exist a network controller or, in the case of asynchronous systems, a network plan. With the latter all units operate without any directions from a central controller. This is the traditional communication scheme, and network control is carried out by instructions given to the system controller/operator prior to operation. Network control is dedicated to the task of maintaining the signal flow in the network in the presence of jamming and eavesdropping, where diverging interests exist between the individual unit and the network. As an example, increased activity of one unit to resist jamming may cause self-jamming to other links (Chapter 6) in the same net through an escalation of transmission power, and thus make eavesdropping simpler.

The extent of network control varies substantially from system to system, and will have to be coupled to operational demands and plans. Different scenarios will also require changes in the network plan, as the same units will have to be used at different geographic locations.

In this chapter, attention will be focused on the challenge of coding and decoding data; this is generally accepted as the basic problem in implementation.

4.2 Signal processing technology

Before discussion of the wide field of component technology, a brief comment should be given on the basic mathematics involved. Even the very complex algorithms can be broken down into a succession of basic operations – multiplications, additions and, on a higher level, correlations and Fourier transforms.

A time series $a(t)$ has a Fourier transform

$$A(f) = \sum_t a(t) \exp(j2\pi ft)$$

The power spectrum of $a(t)$ can be obtained either by squaring the amplitude spectrum $A(f)$

$$S(f) = |A(f)|^2$$

or by taking the Fourier transform of the autocorrelation function of $a(t)$, $R(\tau)$. $R(\tau)$ is given as

$$R(\tau) = \sum_t a(t)a(t - \tau)$$

This is illustrated in Fig. 4.2.

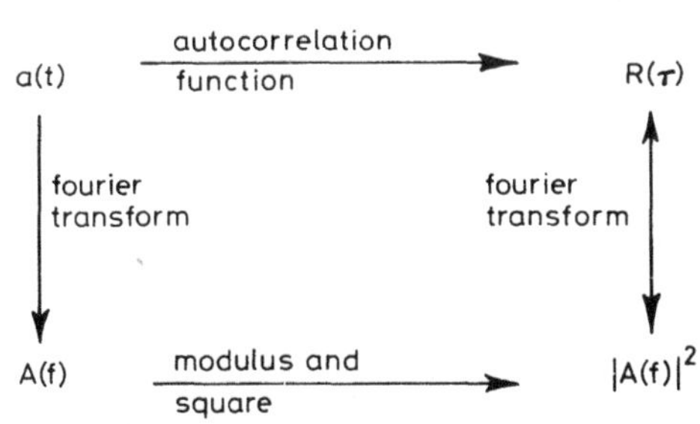

Fig. 4.2 *Connection between time and frequency domain operations*

Given two time series, $a(t)$ and $d(t)$, the corresponding frequency spectra $A(f)$ and $D(f)$, we easily find that the Fourier transform of the product time series $a(t)d(t)$ can be found by convolution (denoted by $*$) of the two frequency spectra $A(f)$ and $D(f)$:

$$\mathrm{FT}\{a(t)d(t)\} = A(f) * D(f)$$

There is a very important point concerning the correlation of two time series. We find that the usual correlation of $a(t)$ and $d(t)$ is

$$R_{ad}(\tau) = a(t) * d(t) = \sum a(t)d(t - \tau)$$

The Fourier transform of this is

$$\mathrm{FT}\{a(t) * d(t)\} = \mathrm{FT}\{R_{ad}(\tau)\} = A(f)D(f)$$

(It will later be shown how correlation can be performed using digitized data.) However, for some analogue processing devices the time axis will have to be inverted because of physical phenomena. The result of correlating two signals where one, $d(t)$, has a reversed reference time axis is

$$C_{ad}(\tau) = a(t) * d(-t) = \sum a(t)d(\tau - t)$$

From Fig. 4.3 the Fourier transform of this is

$$\mathrm{FT}\{a(t) * d(-t)\} = A(f)D^*(f)$$

where the asterisk denotes complex conjugate.

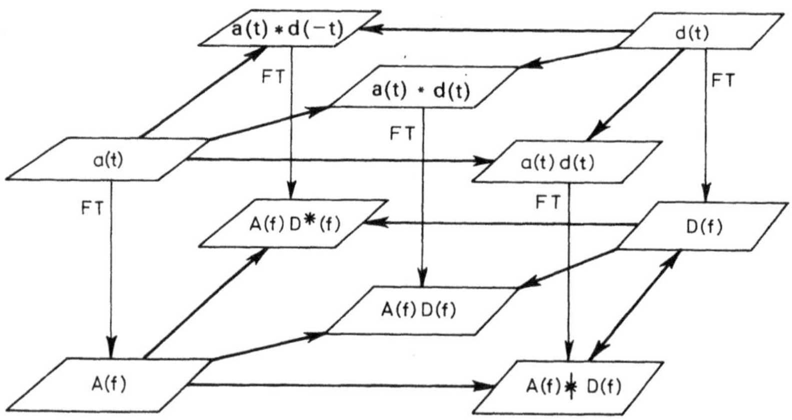

Fig. 4.3 *Connection between time and frequency operations of the two time series a(t) and d(t)*

This distinction is very essential when it comes to implementation, as the inverted moving time reference of analogue convolvers imposes special requirements on the supporting hardware.

Devices can be used that have no memory, such as addition and multiplication units, or networks or semiconductors with no induction, capacitance or time lag. It is seen from Fig. 4.3 that convolution or correlation in the frequency domain is provided by using devices with no memory in the time domain, and vice versa.

Memory devices will be needed for such operations as

Convolution, correlation
Fourier transformation
Filtering in frequency domain.

The optimum use of technology can be achieved by selecting time domain devices for operations best performed in the time domain, and frequency domain units for frequency domain operations.

4.3 Design of spread spectrum communication systems using analogue processing devices

4.3.1 *Surface acoustic wave (SAW) convolver*

In spread spectrum communication, surface acoustic wave (SAW) devices have become of interest mainly because of the very high bandwidth achievable. The devices are basically analogue delay lines with a propagation delay of a few microseconds per centimetre, an operating frequency in the range 1–1000 MHz and fractional bandwidths up to 100%.

When fixed function matched filters are used as a processing device in direct sequence spread spectrum communication, their usefulness is limited because repeated use of the same spreading code makes the system increasingly vulnerable to eavesdropping and intelligent jamming. Thus there is an increasing interest in a SAW processor with variable parameters in the form of a programmed matched filter. This may be implemented in a SAW processor performing the convolution of signals against a variable reference function.

The basic concept of acoustic convolvers is to launch acoustic waves on opposite ends of a piezoelectric crystal with sufficient acoustic amplitude to activate nonlinear effects in the crystal. As the two waves overlap, they will interact through the nonlinearities to induce a potential on the surface of the crystal which is proportional to the product of their amplitudes. If this potential is integrated along the surface, the total voltage is proportional to the convolution between the signal leaving the two transducers. It is, however, necessary for one of the input signals to be time inverted in order to compensate for the effect of wave propagation in opposite directions.

If the two SAWs with wave vectors $\boldsymbol{k}_1$, $\boldsymbol{k}_2$ and frequencies f_1, f_2 are launched in opposite directions ($\boldsymbol{k}_1 = -\boldsymbol{k}_2$) and with the same frequencies ($f_1 = f_2$), the resultant product wave will have wave vector $\boldsymbol{k} = \boldsymbol{k}_1 + \boldsymbol{k}_2 = 0$ and frequency $f_1 + f_2 = 2f_1$. The convolver is then said to be degenerate, and the product wave can be picked up by integrating the potential along the convolver. If $\boldsymbol{k}$ were nonzero, an interdigital transducer would be needed to couple out the convolved signal.

One class of SAW convolver, which does not require expensive and sophisticated airgap supports, is the acoustoelectricplate class of consolvers, in which polarization due to the surface acoustic waves induces charges on a metallization evaporated on to the piezoelectric material.

Fig. 4.4 shows an example of a degenerate convolver design [1]. The convolver is to be used in a spread spectrum system where data demodulation is performed by differential phase comparisons of successive data bits (DPSK).

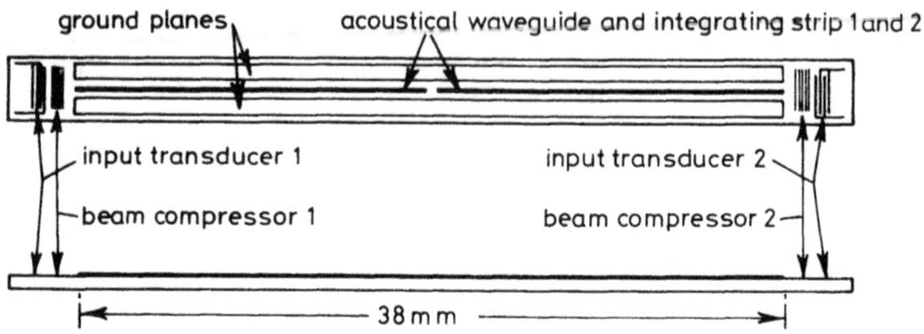

Fig. 4.4 *Reported convolver*
(Reprinted from 'Fast synchronisation modem for spread-spectrum communication system using burst-format message signalling', *IEE Proceedings F*, vol. 128, no. 6, pp. 370–378, November, 1981.)

The nonlinearities generating the product signal are a combination of the elastic, piezoelectric, electrostrictive and dielectric properties of the crystal [2]. Convolving efficiency is dependent on crystal orientation relative to the direction of SAW propagation. Optimum efficiency is obtained by a *YZ* cut on $LiNbO_3$, and in this orientation the interaction is dominated by the third-order piezoelectric nonlinearity (e_{12313}).

The application of this acoustic waveguide convolver as a processing element in a spread spectrum communication system will be treated in the next section.

4.3.2 *Application of SAW convolvers to spread spectrum burst communication systems*

In this section a spread spectrum communication system incorporating minimum shift keying (MSK) spreading modulation and DPSK data modulation is discussed in terms of the SAW convolvers introduced in Section 4.3.1.

To lessen the dangers of eavesdropping and intelligent jamming during a data transmission burst, a long code is segmented and used for spectrum spreading. As each data bit is DPSK modulated and uses consecutive and different segments of a longer code for spreading, two spreading code segments must be serially loaded for each data bit detection. Thus a continuous data demodulation requires two DPSK convolvers operated in a bit staggered fashion. The two convolvers must be able to store 2 × (spreading code chip rate/data bit rate) = 2 × 1023 chips, equivalent to two data bits, and thus two electrically isolated substrates constitute one DPSK convolver as shown in Fig. 4.4. Fig. 4.5 shows the operation of the demodulation at one instant. D_i denote the modulated data bits and K_i denote the spreading codes, where the overbar indicates a time reversed image. X is a don't care situation. At the instant shown the first data bit is detected in convolver 1.

The use of a SAW processor as a convolver requires synchronization algorithms to align received signal and reference, but the time allowed to synchronize must usually be kept short in order to exploit the advantages of large bandwidths and possible fast burst message transfer. This requires a suitable preamble. To use segments from a longer sequence is not desirable because it is possible that large sidelobes due to unbalanced segments may upset the synchronization process (see Chapter 3). The preamble chosen is a maximal length sequence with a period of 2047, matching the duration of two data bits and thus the DPSK convolver. The preamble is not modulated by data and the detection concept presupposes random arrival time of the message signal.

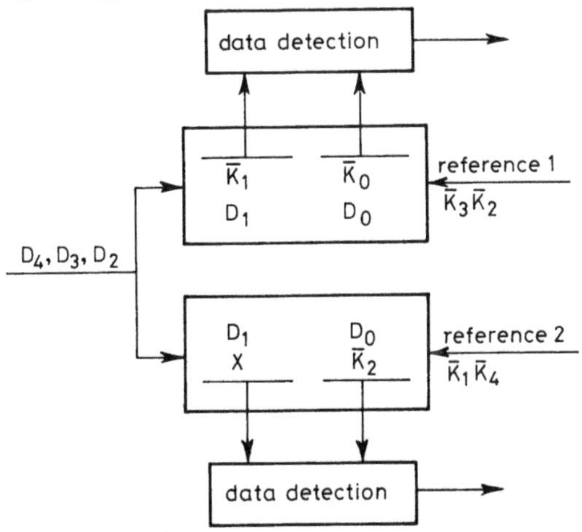

Fig. 4.5 *DPSK data demodulation using two convolvers*
(Reprinted from 'Fast synchronisation modem for spread-spectrum communication system using burst-format message signalling', *IEE Proceedings F*, vol. 128, no. 6, pp. 370–378, November, 1981.)

Coarse synchronization of convolver reference signal and received signal under these conditions introduces special and significant problems related to correlation ambiguities. The convolver differs from a usual matched filter in the sense that the reference is not merely stored but rather recirculated. Thus the received signal and the reference may produce not only one correlation peak at the output for each data bit period but also multiple peaks. The peaks differ considerably in magnitude and it is important to choose the largest possible. However, the problem is that, at the instant when one correlation peak is detected, the receiver has no information on whether a larger peak will arrive later or if the largest peak has been missed owing to the random phase of the recirculating code reference when the signal arrived.

For the reported design the two DPSK convolvers are linked to a pair of code generators recirculating code references which are time reversed images of the expected transmitted preamble.

The two references are identical but are time displaced by one time bit. The references and the received signal propagate in opposite directions in the convolvers so that they have an apparent relative speed twice the actual speed of either. It takes four bit times for a received preamble to pass completely through the convolver, and thus four partial correlation peaks, spaced one bit time apart, will result at the output of each convolver. Because the references in convolvers 1 and 2 are displaced by one time bit, and because the references and received signal have an apparent relative motion of twice the actual motion, the correlation events are displaced by half a time bit at the two convolver outputs. The events are illustrated in Figs. 4.6 and 4.7 using a coordinate system of time, phase and correlation peak amplitude. The figures show that the four partial correlations occur regardless of the received signal arrival time. Note that there exists one case when only three partial correlations occur. This happens when the second correlation reaches maximum value. As the preamble propagates through the convolver, the maximum correlation possible increases from zero to full and then back to zero in a triangular fashion.

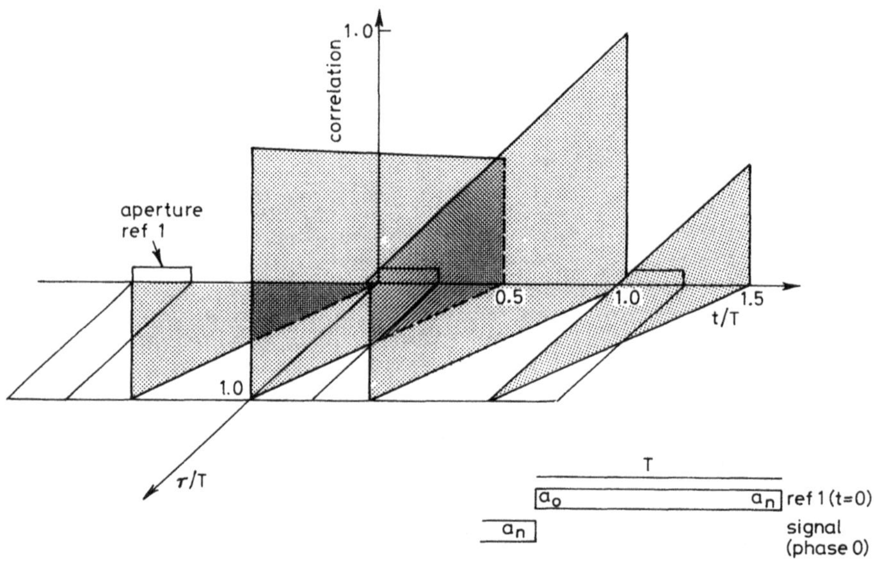

Fig. 4.6 *Correlation function for reference 1 and received signal. t/T is the time axis and τ/T represents the phase of the received signal*
(Reprinted from 'Fast synchronisation modem for spread-spectrum communication system using burst-format message signalling', *IEE Proceedings F*, vol. 128, no. 6, pp. 370–378, November, 1981.)

When the reference is displaced relative to the convolver interaction region and partial correlation with the received signal occurs, the displacement is called *misalignment*. Thus the multiple partial correlations introduce ambiguities and detection methods are required that can guarantee that only the largest peak produces a synchronization. Some systems [3] use coincidence detection and logic gating to remove the ambiguities and guarantee correct synchronization. This method usually requires storing and delay or the output from one of the convolvers until there is a decisive output from the second convolver. The practical realization of delay and store is considered nontrivial, and the coincidence method may also require a nonprocessed or dummy bit in the message preamble to align local reference and received signals, which increases the time to synchronize. These problems are avoided using a scheme of time displaced references and aperture signals.

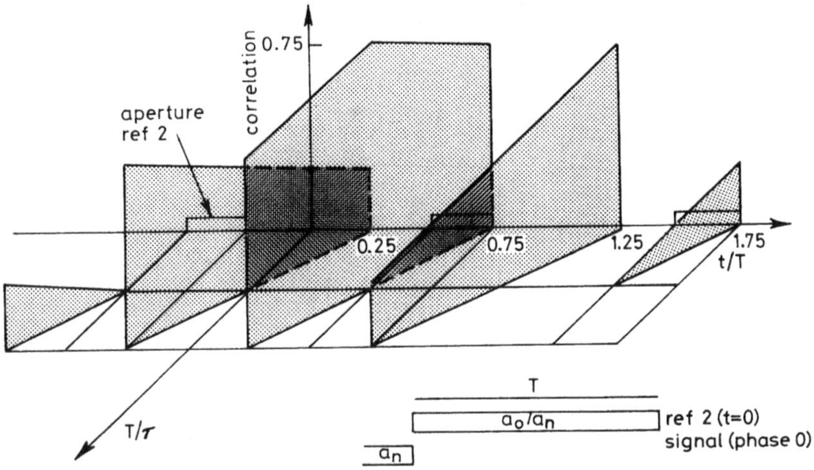

Fig. 4.7 *Correlation function for reference 2 and received signal. t/T is the time axis and τ/T represents the phase of the received signal*
(Reprinted from 'Fast synchronisation modem for spread-spectrum communication system using burst-format message signalling', *IEE Proceedings F*, vol. 128, no. 6, pp. 370–378, November, 1981.)

The synchronized aperture signals are simple aperture window functions defining what may be said to be useful regions of correlation. There are two window functions, one for each of the DPSK convolvers, and they open at time intervals displaced by half a bit time. The functions are traced in Figs. 4.6 and 4.7. They remain open for half a bit time every two bit times. With a 25% duty cycle for each aperture window and the use of two convolvers recirculating the same, but time displaced, sequence, the received signal is observed half the time. This is, however, acceptable because of the time compression due to counter-propagating references and received signal. The time–bandwidth product of input and output signals remains the same, and no significant information is removed.

The correlation at any instant is defined as

$$CORR = (S_l) \cap (A^j) \qquad 0 < l < p - 1 \qquad j = 1, 2$$

where S is the received sequence of elements $a_0, a_1, a_2, \ldots, a_{2046}$, and $A^j = \bar{S}$ is the time-displaced reference, the overbar denoting time reversal. The parameter l describes the distance from the edge of the output transducer to the position of the received signal.

Fig. 4.8 is a simple time/position diagram and shows correlation events in convolver 1 and convolver 2 for a particular phase when the preamble will be detected in convolver 1.

The aperture window for convolver 1 opens after a loading of the reference so that its starting point is at the edge of the output transducer. The aperture window remains open for half a bit time and thus closes before the first partial correlation occurs in convolver 1. The aperture window for convolver 2 opens after a loading of the reference so that its starting point is at the edge of the

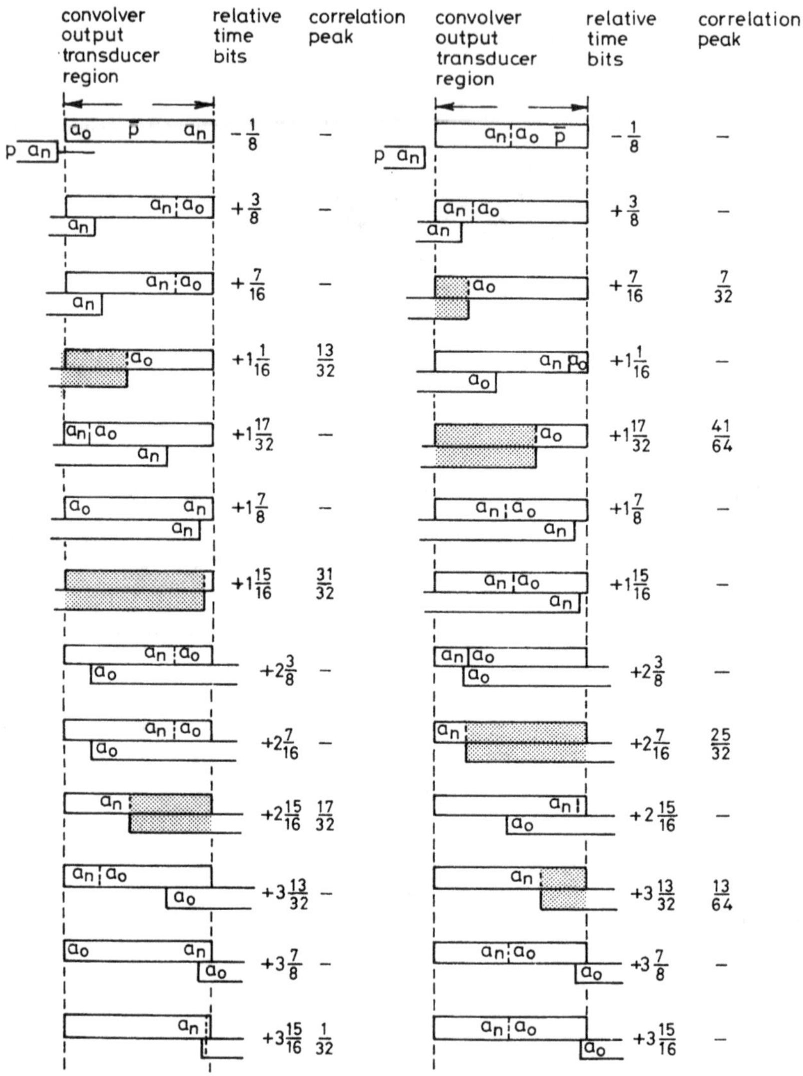

Fig. 4.8 *Time/position diagram for convolvers 1 and 2 showing typical case of partial correlation during synchronization*

output transducer, as was the case for convolver 1. Since the references in convolvers 1 and 2 are time displaced by one bit time, the aperture window in convolver 2 opens one bit time displaced from the aperture window in convolver 1.

Thus the window for convolver 2 is closed during the first partial correlation.

Since two partial correlations occur one bit time apart during each reference recirculation, not more than one partial correlation peak is observed in the window time interval. As Fig. 4.8 shows, the aperture window for convolver 1 is chosen in such a way that, during the first open interval after the preamble is received, at least three quarters of the preamble chips are in agreement with the circulating reference. This is for the case when the preamble is detected by convolver 1. In Fig. 4.8, the preamble is detected with a correlation peak of 31/32 of the maximum possible value.

The aperture window in convolver 2 is chosen in the same way as for convolver 1, and thus, for the situation pictured, the second partial correlation in convolver 2 occurs when the window is closed.

It takes four bit times for a received preamble to pass completely through the convolver, and thus another partial correlation peak is possible during the second interval of two bit times while the window is open. However, the third partial correlation in convolver 1 takes place when the aperture window is closed.

During the second open interval, a maximum of one quarter of the preamble sequence chips in the received signal is in agreement with the reference. For the case in Fig. 4.8, the correlation peak is 1/32 of the maximum possible value. Thus the largest of the two partial correlation peaks always occurs first with a value of at least three quarters of the maximum possible value.

In the situation in Fig. 4.8, the third and fourth partial correlations in convolver 2 occur when the window is closed. Thus, in this case, the preamble is not detected by convolver 2 at all. However, when the signal is not detected in convolver 1, it is detected by convolver 2.

Thus observation of the signal for the complete time T shows a small and unwanted partial correlation peak in either convolver 1 or 2 during later periods. However, they are not of much concern since they occur later than the desired peak. Fig. 4.9 shows the phase/time/correlation function for a randomly arriving message as seen through the aperture windows of convolvers 1 and 2. As the figure points out, the largest partial correlation peak is seen first, either by convolver 1 or by convolver 2. Thus there is no need for storing, delaying and logic gating to determine the received signal arrival time. To bring the received signal and references in bit alignment for full correlation, the receiver uses counters that keep track of the recirculating references. The use of time displaced references and two aperture window functions ensures that the largest partial correlation peak is detected within an epoch of two time bits from the appearance of a message. This is important because, after the two time bits of the preamble, the next bit in the message will enter the convolver, and, in order to perform demodulation, the references must also start to enter the convolvers at this time. Thus the signal processing used avoids the need for nonprocessed or dummy bits in the message.

If the peak is detected before two time bits have elapsed, the code generators are stopped, loaded with their starting vector and then restarted on information from the counters. The counters keep track of the reference time base and determine by how much it must be shifted to achieve perfect alignment. This removes any loss due to misalignment.

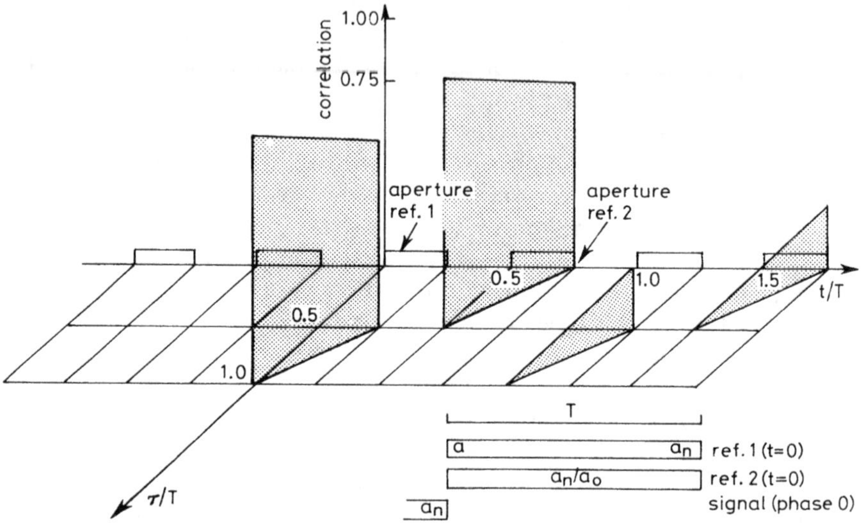

Fig. 4.9 *Correlation function for random arriving preamble as seen through the aperture windows of convolver 1 and convolver 2. t/T is the time axis and τ/T represents the phase of the received signal*
(Reprinted from 'Fast synchronisation modem for spread-spectrum communication system using burst-format message signalling', *IEE Proceedings F*, vol. 128, no. 6, pp. 370–378, November, 1981.)

The modulator assembly used for the IF subsystem consists of a data source running at a rate of 90.420 kHz followed by a differential phase shift modulator. This data sequence is spread by a 1023 bit code which is running synchronously with the data source. The spreading sequence has a chip rate of 92.500 MHz. Code generation and spreading modulation is implemented by emitter coupled logic (ECL) components with rise time and delays of 1–2 ns.

The choice of RF modulation scheme is crucial to the system performance. Military tactical communication requires maximal spectral flatness and constant envelope waveform in order to minimize jamming and intercept susceptibility. Phase or frequency shift keying gives the required constant envelope, but out-of-band energy is high owing to the sharp transitions in the RF waveform. Minimum frequency shift keying (MSK), as used in this system, is more advantageous than PSK or FSK because fractional out-of-band power is reduced and the signal waveform has a continuous first derivative. MSK is performed as depicted in Fig. 4.10.

The modulating code chip stream is converted to an IF waveform by hopping between two frequencies f_1 and f_2 separated by half the chip rate. This ensures that the radio waveform will have exactly one half of a period extra at frequency

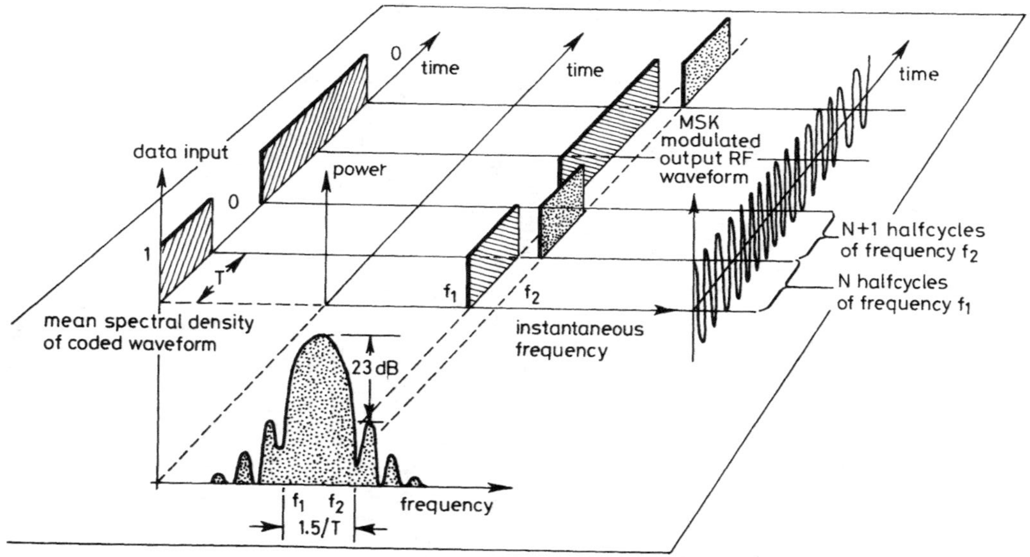

Fig. 4.10 ***MSK modulation scheme of order M = 6***
(Reprinted from 'Fast synchronisation modem for spread-spectrum communication system using burst-format message signalling', *IEE Proceedings F*, vol. 128, no. 6, pp. 370–378, November, 1981.)

f_2 compared with the period at frequency f_1. By switching frequency at the top of the waveform, the first derivative will be continuous and the out-of-band power and spectral rolloff is at an optimum [4]. Fig. 4.10 also shows mean spectral energy for an MSK modulated pseudonoise code. The spectrum has a mainlobe of width $1.5/T_c$, where T_c is the chip duration, and has a first sidelobe 23 dB down, compared with 13 dB for PSK and FSK. It has been reported that out-of-band power can be further reduced not by making the frequency hopping instantaneous but rather by employing successive transfer between the two frequencies.

In the reported design, f_1 is chosen as 277.500 MHz and f_2 as 323.750 MHz, giving a null-to-null bandwidth of 138.750 MHz matched to the convolver waveguide. This corresponds to six half cycles of frequency f_1 or seven half cycles of frequency f_2 in each code chip.

MSK signalling is implemented as proposed by Amoroso and Kivett [5]. This method implies that a PSK modulated carrier of frequency f_2 is passed through a bandpass filter with sin x/x transfer function centred at frequency f_2 and a null-to-null bandwidth equal to that of the PSK signal ($2/T_c$ = 185.000 MHz) [6]. The PSK modulation of this scheme is critical with regard to transition time and signal delays, but this has been solved by means of ECL and hot carrier diode double balanced mixers. The filter has been implemented by SAW technology.

Measurements have been performed to evaluate the modem performance in wideband and narrowband jammer environments during forced synchronization.

The processing gain offered by the spreading code will, in principle, give equal protection against wideband and narrowband jammers of equal energy within the processed frequency band. This system with a spreading code chip rate/data bit rate of ~1000 should thus allow for jammer energy to exceed signal energy by up to 30 dB before detection becomes impossible.

Extensive tests have been performed with CW jamming of the transmissions. The frequency spectrum of the transmitted code consists of lines separated by the inverse of the code length, and the energy in each line is given by the MSK power distribution. A CW jammer will beat with each of these lines, and the sum of all these beats will be seen as sidelobes in the correlator output signal. The sidelobe pattern will have a rapid frequency dependence as the beat frequencies change, but the mean value will remain constant. The detectability of the spread spectrum signal should thus be unaffected by the jammer down to a 30 dB negative signal/jammer level ratio, where the sidelobe level approaches that of the correlation peak. Experiments have verified that correct operation is attainable down to −28 dB *SNR*. The degradation from 30 to 28 dB is due to internal reflections within the convolver, where the coded signal beats with its time symmetrical image and raises the overall peak sidelobe level by 2 dB.

Experiments have also verified that pulsed CW jamming gives no further degradation relative to continuous jamming even if the pulse frequency is phased to the code chip frequency.

If wideband white noise jamming is applied, the obtainable jammer spectral intensity will generally be much lower. The threshold setting of the correlation peak detection circuit will have to be adapted to a maximum allowable false alarm rate. This is the same situation as in pulsed radar systems, where a target (correlation) is to be detected in the presence of clutter (noise), and the same theory applies.

It is important to note that, owing to the expected stochastic noise in a practical system, the probability of detection will never equal unity because a high probability of detection corresponds to a high rate of false alarms. Assuming white Gaussian noise, 99% probability of detection is obtained at a +15 dB signal/noise level with a threshold setting corresponding to a false alarm rate of 10^{-8}. The 1023 chip code improves this figure by 30 dB, thereby giving a 99% probability of detection at a 15 dB negative signal/noise ratio.

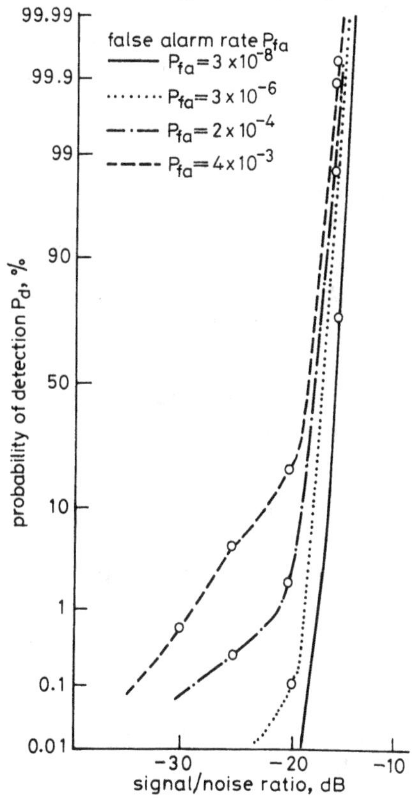

Fig. 4.11 ***Experimental performance of synchronized modem with white Gaussian jammer*** (Reprinted from 'Fast synchronisation modem for spread-spectrum communication system using burst-format message signalling', *IEE Proceedings F*, vol. 128, no. 6, pp. 370–378, November, 1981.)

Experiments have confirmed these data. The performance of the system with respect to wideband Gaussian noise has been measured to a precision of 10^{-6} (10^{-8} samples). Fig. 4.11 shows the probability of detection against signal/noise ratio, with the false alarm probability (threshold setting) as the parameter. These curves give the basic performance of the system. They are within the significance of the measurements in accordance with theory. As can be seen from the curves, the probability of detection exceeds 99.99% for signal/noise ratios greater than -13 dB at false alarm rates of less than 10^{-8}.

At the 99% level of detection, which is a reasonable point of operation during transmission, the necessary input level varies by less than 2 dB for false alarm rates between 10^{-3} and 10^{-8}. During synchronization, a successive reduction in the threshold setting and corresponding increase in the false alarm rate is performed until synchronization is acquired. The present starting threshold is high when the signal/noise ratio was high during the previous transmission; otherwise the threshold is set to the allowable minimum given by the maximum false alarm rate.

The conclusion of the performance evaluation is that the modem performs according to expectations. The protection given by the spreading code is maintained, even during severe jamming conditions, at the ultimate chip rate of 92.5 MHz.

4.4 Design of spread spectrum communication systems using digital processing devices

4.4.1 *Technology of digital processing devices*

There are two main types of circuit in semiconductor integrated circuit technology:

Bipolar technology (TTL, ECL, SOSMOS etc.)
Unipolar technology (NMOS, CMOS etc.).

Traditionally bipolar technology has been faster, giving the designer a considerable ability to design relatively complex and fast circuits. There are several families of bipolar technology and new ones are still appearing. Most of the families in this group have to use high power consumption to achieve the high speed.

Compared with MOS technologies the bipolar technologies are difficult to use in a full custom design to realize unusual and special architecture. To overcome this problem, a lot of bipolar gate arrays are now obtainable to implement systems of medium complexity. As described later, the customer then only has to specify the metal interconnection on prefabricated wafers.

4.4.2 *Design of very large scale integrated circuits*

The designer of very large scale integrated circuits (VLSIs) needs to be familiar with the signal processing algorithms as well as the integrated circuit procedures. There exist various philosophies, and the choice of which to use should be carefully considered in advance of the construction.

In the *gate array* approach, only one layer is designed. This is an approach which is suitable for industry, as the startup cost and turnaround are low. Many companies offer this possibility, and computer aided design (CAD) machines are readily available. The reliability of this process is very satisfactory.

Another approach is to provide the designer with a standard set of processing elements (cells). The designer then connects cells to build his dedicated processor. This is the *standard cell* approach.

Full *custom design* of integrated circuits requires CAD tools at every step in the design. The CAD machines can be provided by a design centre or may be owned by the designer himself. The basic philosophy is that the person who knows the problem designs the circuits. Very interesting and effective solutions of processing tasks are possible, and this approach is increasing in popularity as the cost of CAD tools decreases.

Faster and more effective systems cannot be obtained merely by increasing the speed for digital integrated circuits. As a matter of fact, the speed is limited by the electrical signals transmitted on wires. The organization or structure is of special importance. In a von Neumann machine the instructions are sequentially fetched and executed. A computer system built around such a computer will never be faster than the bandwidth of the memory connected to it.

With VLSI it is quite easy to use a lot of simpler processors for solving a given problem. The total speed is thereby enhanced using parallel processing. The question is one of how they should be connected to achieve a high capacity system.

One type of parallel machine is the multiprocessor consisting of a lot of separate processors connected together, sharing the load and exchanging data.

Quite often it would be a good idea to execute as many operations as possible each time a data element has been fetched from the memory. Two types of architecture look specially interesting in this conjunction: pipeline and systolic array.

A *pipeline* can be thought of as consisting of a series of working cells where each unit performs a specific task on all objects passing. Pipeline is a system where data is shifted through a linear array of processors. The optimal flow of data is achieved when the processing time in each element is comparable.

A *systolic array* consists of a lot of regularly connected cells, each capable of performing operations. A primary feature of each array is the simple, regular data flow, together with simple control of the cell functions. Transmission of information inside the array is between neighbouring cells. Only cells in the periphery communicate with memory. This architecture is especially suitable when many operations have to be performed on each element in a large data memory. The performance in a systolic array is due to the parallelism of many simple cells and not the sequential use of a few, powerful processors as found in some conventional architectures.

This regular structure of simple cells makes the system especially suitable for VLSI implementations. This is because it is necessary to design and test only a few, simple cell types. An efficient VLSI architecture has the following features:

1 The architecture can be implemented using a few types of simple cells.

2 The architecture has a simple and regular data flow and control structure. The cells can then be put together in a network by using local connections. The long distance interconnections or non-regular communication is kept at a minimum.
3 The architecture makes extensive use of the pipeline principle and parallel processing. This is achieved when a great number of cells are active at the same time. The overall performance is therefore good.

The degree of parallelism in a specialized system is determined by the underlying algorithm. A high degree of parallelism is only achieved if the architecture meets the specifications listed. This means that a good algorithm for VLSI implementation is not necessarily the one that gives a minimum of computations. The most expensive factor in VLSI realizations is communication, not processing power.

4.4.3 *Case study: design of LSI digital correlator*

This section reports the design of an LSI correlator in CMOS/SOS technology featuring high bit rate and low power consumption. A simplified block diagram of the correlator is shown in Fig. 4.12. The main components are three 64 bit registers, 64 exclusive-OR gates and 64 current generators which can be turned on and off individually. Two of the registers have serial input but all have parallel inputs.

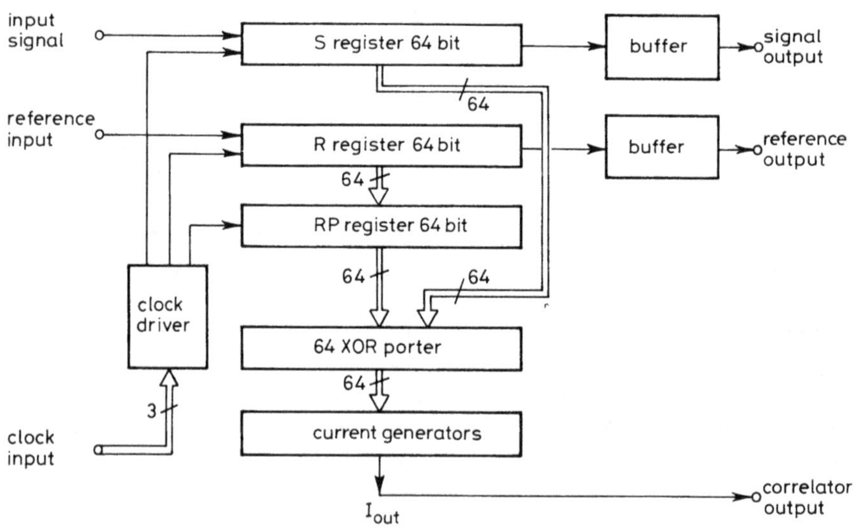

Fig. 4.12 *Block diagram for 64 bit correlator*

One of the serial input registers acts as a delay line for the input signal and the other is used to store the reference signal. In order to allow for cascading, it is possible to pass on the reference and input signals from the end of one correlator through buffer circuits to another correlator chip. It is also possible to copy from the reference register to the parallel in/parallel out register. This will allow change of reference code with no delay in data demodulation.

Correlation is executed by comparing the input signal and the reference signal in 64 exclusive-OR ports. The output from these ports will be 0 if the inputs agree and 1 if they do not agree. The output from the ports will switch current generators on or off. The currents I_0 are summed on a bus and the sum will be proportional to the degree of correlation:

$$I(n) = I_0 \sum_{i=0}^{p} \theta'(i, n) \tag{4.1}$$

$$\theta'(i, n) = \begin{cases} 1 & a(i) = r(n - i) \\ 0 & \text{otherwise} \end{cases}$$

$a(i)$ is the code reference in bit position i and $r(n - i)$ is the received signal after the ith stage of the delay line at time $n(1/\text{bit rate}) = nT_c$.

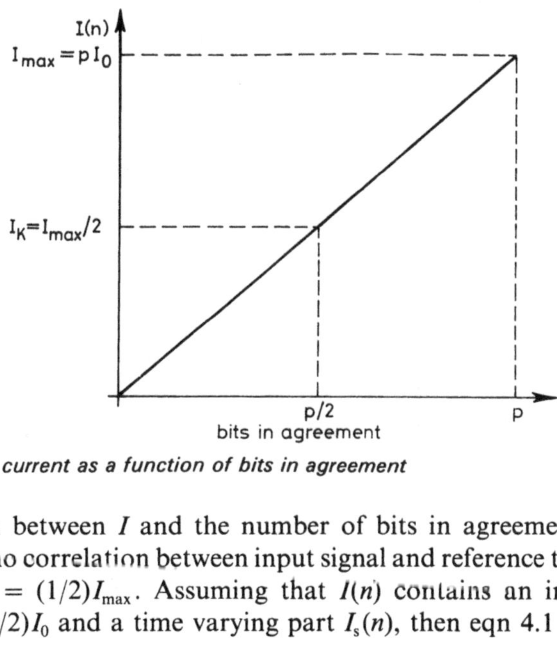

Fig. 4.13 *Output current as a function of bits in agreement*

The relation between I and the number of bits in agreement is shown in Fig. 4.13. For no correlation between input signal and reference the value of $I(n)$ will be $(p/2)I_0 = (1/2)I_{max}$. Assuming that $I(n)$ contains an invariable component $I_c = (p/2)I_0$ and a time varying part $I_s(n)$, then eqn 4.1 can be written as

$$\begin{aligned} I(n) &= I_c + I_s(n) \\ &= \frac{p}{2} I_0 + \frac{I_0}{2} \sum_{i=0}^{p} \theta(i, n) \end{aligned} \tag{4.2}$$

$$\theta(i, n) \begin{cases} = +1 & \text{if } a(i) = r(n - i) \\ = -1 & \text{otherwise} \end{cases}$$

Mapping the additive group with 1 and 0 to the multiplicative group with -1 and $+1$ allows $\theta(i, n)$ to be expressed as a multiplication between $a(i)$ and $r(n - i)$:

$$I(n) = \frac{p}{2} I_0 + \frac{I_0}{2} \sum_{i=0}^{p} a(i) r(n - i) \tag{4.3}$$

Eqn 4.3 shows that the correlator can be viewed as a linear filter for signals being binary quantized. The impulse response is equal to the stored reference and is also binary.

Even though it is only possible to correlate binary signals in each correlator it is feasible to connect M correlators in 'parallel' and then perform correlation on input signals quantized to 2^M levels. The input signal $r(n)$ can then be expressed as

$$r(n) = \sum_{m=0}^{M-1} r_m(n) 2^{-m} \tag{4.4}$$

$r_m(n)$ is digit m with weight 2^{-m} taking on values ± 1.

Inserting eqn 4.4 into eqn 4.3 and changing the order of summation gives

$$I(n) = \frac{p}{2} I_0 + \frac{I_0}{2} \sum_{m=0}^{M-1} 2^{-m} \sum_{i=0}^{p} a(i) r(n - i) \tag{4.5}$$

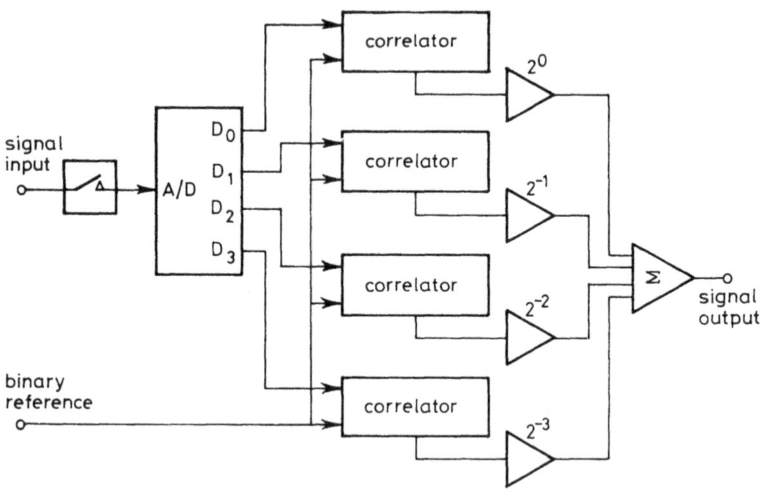

Fig. 4.14 *Realization of multibit quantization*

This equation shows that each bit in the weight may be correlated in separate correlators with the same binary reference. The partial correlations must finally be weighted with a factor 2^{-m} and then summed. Fig. 4.14 shows the realization.

As an example of a design using the silicon on sapphire technology, Fig. 4.15 shows a complete 64 bit correlator packed within 4.9 mm × 2.9 mm. Figs 4.16

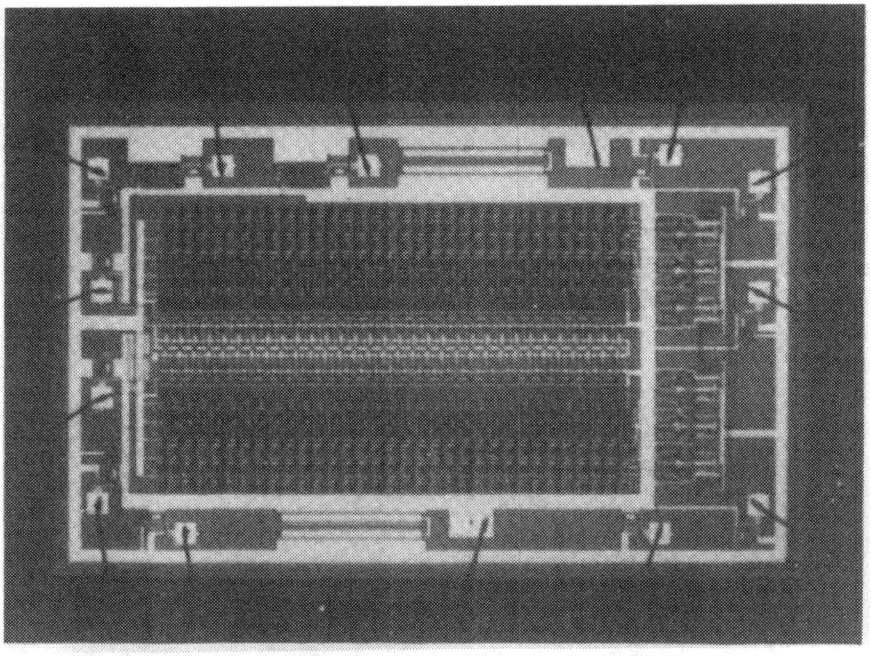

Fig. 4.15 *CMOS/SOS correlator design*

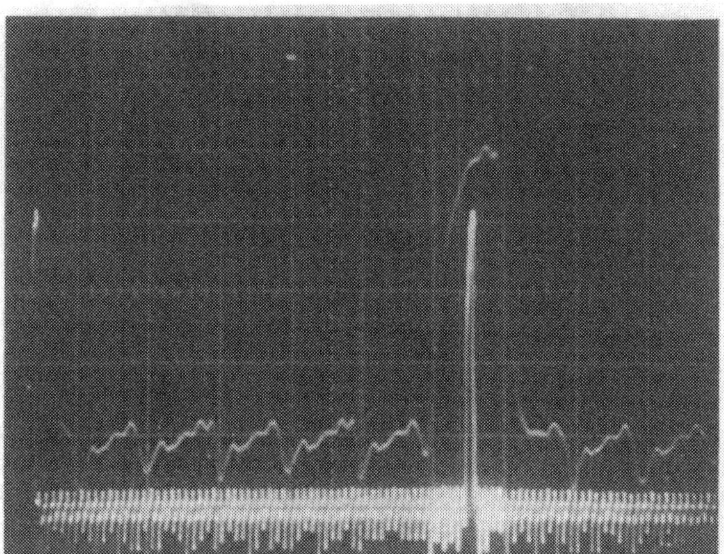

Fig. 4.16 *Correlation output for CMOS/SOS correlator clocked at 5 MHz.*
Lower trace 50 mV/div, 2 μs/div
Upper trace 50 mV/div, 200 ns/div

and 4.17 show the actual correlator performance. The circuit is dissipating 15 mW at 5 MHz and 27 mW at 10 MHz.

4.4.4 *Case study: CCD programmable matched filter*

The digital correlators suffer from the disadvantage that when they are used to process analogue signals – as is the case in spread spectrum communication applications – A/D conversion is required. Realizing analogue processors by using charge coupled devices (CCD) allows analogue input signals, and these memory devices also inherently provide the necessary shift facility. However the provision of an analogue reference in the matched filter necessitates refreshing since the CCD process does not allow long term storage of analogue information.

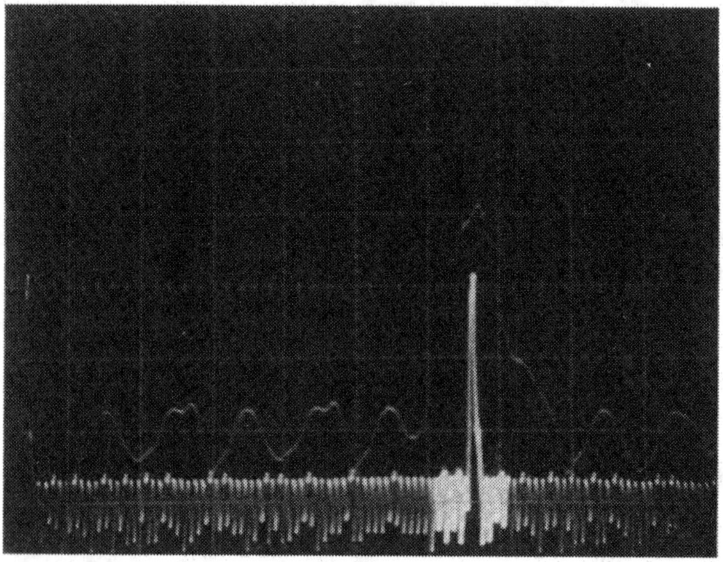

Fig. 4.17 ***Correlator output for CMOS/SOS correlator clocked at 10 MHz.***
Lower trace 50 mV/div, 1 μs/div
Upper trace 50 mV/div, 100 ns/div

A possible formation for matched filter application is to have an analogue signal input but a digital reference. An implementation used in commercially available circuits is shown schematically in Fig. 4.18. The analogue input signal is sampled at a rate twice the clock frequency using a two-phased complementary clock signal. If the samples are numbered continuously, then the odd samples are fed to one of the registers and the even samples to the other.

The contents of each of the two analogue shift registers are then multiplied bit by bit with the content of the binary reference register. These structures do in this way require a multiplying digital/analogue converter at every filter point. The resulting two products are summed and passed every second time to a plus output. A minus output is worked in the same way but now the contents of the binary reference are inverted before multiplication. The final correlation is obtained by subtracting the plus output and the minus output.

One of the fundamental disadvantages with CCD filters is the transfer inefficiency. A small fraction of each charge packet is left behind at each

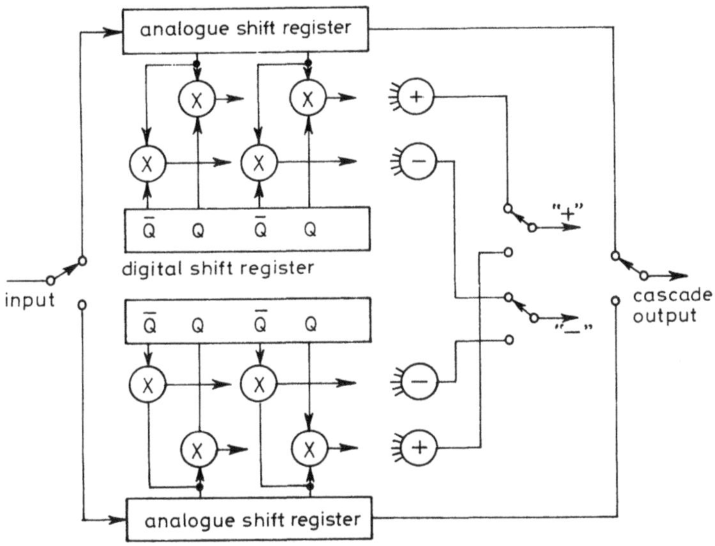

Fig. 4.18 *Schematic operation of a CCD programmable matched filter*

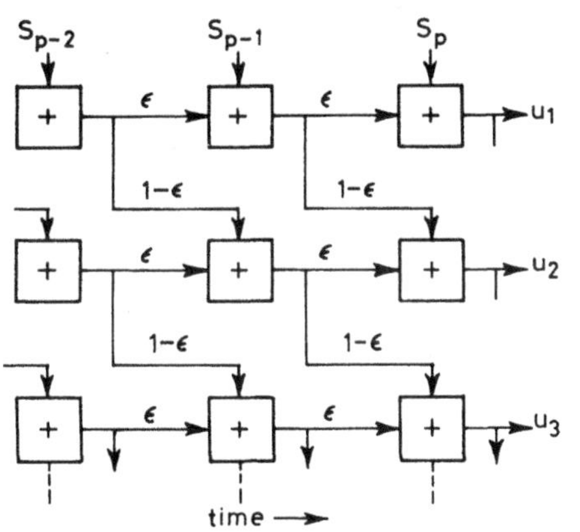

Fig. 4.19 *Operation of the CCD shift registers*

transfer, causing cumulative signal degradation along the register. A parameter ε is usually defined as the ratio between this fraction of charge and the original charge. The situation is illustrated in Fig. 4.19 where S_p, S_{p-1} ... are the individual consecutive samples. The contents of the different tapped stages of the register then become

$$u_1 = S_p + \varepsilon S_{p-1} + \varepsilon^2 S_{p-2} + \varepsilon^3 S_{p-3} + \ldots$$

$$u_2 = (1 - \varepsilon)(S_{p-1} + \varepsilon S_{p-2} + \varepsilon^2 S_{p-3} + \ldots$$

$$u_3 = (1 - \varepsilon)^2(S_{p-2} + \varepsilon S_{p-3} + \varepsilon^2 S_{p-4} + \ldots$$

$$\vdots$$

$$u_k = (1 - \varepsilon)^{k-1} \sum_{i=k}^{\infty} \varepsilon^{i-k} S_{p-i+1}$$

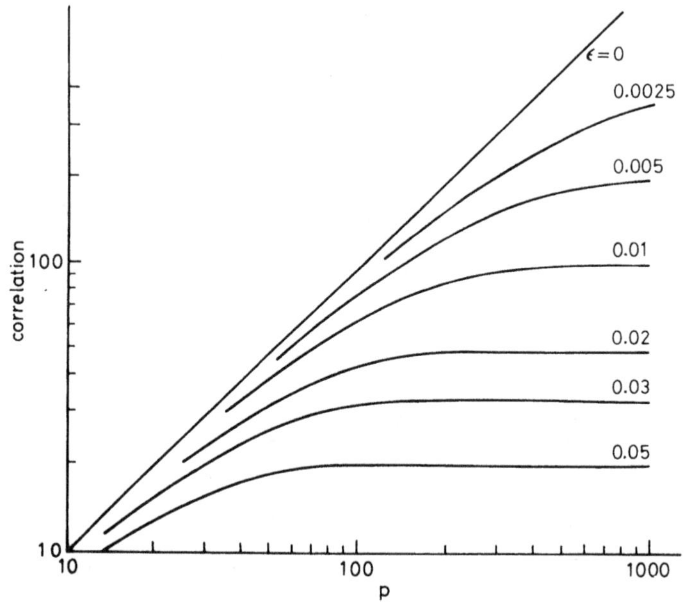

Fig. 4.20 *Correlation peak as a function of shift register length p for selected values of the transfer efficiency*

The maximum value for the peak autocorrelation then becomes

$$R = u_1 S_p + u_2 S_{p-1} + \ldots \quad (4.6)$$

where $|S|$ now are the reference signals. For a p bit filter the correlation peak becomes

$$R = \sum_{k=1}^{p} u_k S_{p-k+1} \quad (4.7)$$

which to first-order approximation gives

$$R \approx S_p^2 + (1-\varepsilon)S_{p-1}^2 + \ldots + (1-\varepsilon)^{p-1}S_1^2 \tag{4.8}$$

For a spreading sequence of ± 1 this gives

$$K \approx 1 + (1-\varepsilon) + \ldots + (1-\varepsilon)^{p-1} = \frac{1}{\varepsilon}[1-(1-\varepsilon)^p] \tag{4.9}$$

A plot of eqn 4.9 is shown in Fig. 4.20 for some values of ε.

The limiting factor which determines the maximum number of possible stages in the filter will be the signal/noise ratio. At some limiting point the addition of an extra stage to the filter will contribute less than the noise level in the filter. Thus no correlation increase will be detected. The increase in correlation output is given by

$$\Delta R = \frac{1}{\varepsilon}[1-(1-\varepsilon)^{p+\Delta p}] - \frac{1}{\varepsilon}[1-(1-\varepsilon)^p] \tag{4.10}$$

which for $\Delta p = 1$ gives

$$\Delta R = (1-\varepsilon)^p$$

Now the filter noise level relative to the signal level is given as $N/S = 10^{-(S/N)\mathrm{dB}/20}$, and if $p_{\max}$ is said to be the number of stages for which ΔR is less than the noise level, then

$$p_{\max} = \frac{\ln(10^{-(S/N)\mathrm{dB}/20})}{\ln(1-\varepsilon)} = 0.115\,\frac{(S/N)\mathrm{dB}}{\varepsilon}$$

if $\ln(1-\varepsilon)$ is approximated by $-\varepsilon$.

Now depending on the filter signal/noise ratio and ε, a maximum number of stages can be operated in series.

The value of ε will depend on the clock frequency. For the commercial available Reticon R5401 binary analogue correlator ε has been calculated to 6×10^{-4} for 1 MHz and $\approx 10^{-2}$ at 2 MHz. This will for practical purposes only allow a relatively short CCD matched filter. However in [7] a general purpose single chip 1024 stage binary/analogue CCD filter is reported. The charge transfer inefficiency in this case was measured to $\approx 10^{-3}$ for a device operating at 1 MHz. The device is designed to eliminate fixed pattern noise due to variations in the tapping transistor characteristic along the length of the filter, and code dependent offset is eliminated using differential channels. It is also possible to compensate for the charge transfer inefficiency effects by programming the device with modified tap weights that invert the dispersion caused by the transfer inefficiency.

References

1 ENGAN, H., INGEBRIGTSEN, K.A., and RØNNEKLEIV, A.: 'Design of SAW convolver for processing MSK modulated waveforms', *Electron. Lett.*, 1980, **16**, (24), pp. 908–909

2 GANGULY, A.K. and DAVIS, K. L.: 'Theory of degenerate SAW elastic convolver', Proceedings of IEEE Ultrasonic Symposium, 1979, pp. 715–724
3 BRODTKORB, D. and LAYNOR, J.E.: 'Fast synchronisation in a spread spectrum system based on acoustoelectric convolver', Proceedings of IEEE Ultrasonics Symposium, 1978, pp. 561–566
4 SIMON, M.: 'A MSK approach to offset QASK', *IEEE Trans.*, 1979, **COM-27**, pp. 921–923
5 AMOROSO, F. and KIVETT, J. A.: 'Simplfied MSK signalling technique', *IEEE Trans.*, 1977, **COM-25**, pp. 433–441
6 SMITH, W.R.: 'SAW filters for CPSM spread spectrum communication', Proceedings of IEEE Ultrasonics Symposium, 1977, pp. 524–528
7 HAKEN, R.A., PETTENGILL, R.C., and HITE, L. R.: 'A general purpose 1024 stage electronically programmable transversal filter', *IEEE J. Solid-State Circuits*, 1980, **SC-15**, (6), pp. 984–996

Chapter 5

Propagation medium

5.1 Introduction

The effect of the propagation environment on spread spectrum modulated signals depends on a number of factors, but may differ considerably from what is usually seen on more conventional narrowband systems.

Generally the received signal may include a direct path component and several multipath signals. The multipath signals may in turn consist of a single specular component, invariant with the exception of possible Doppler, and a diffuse component spread in both time and frequency. Thus, depending on whether or not there is a significant specular multipath component, the envelope of the multipath signal may be Rician or Rayleigh distributed.

For a line-of-sight radio link reflections from terrain or manmade obstacles will result in multipath. Depending on the wavelength of operation and the roughness of the reflectors, the multipaths will be either specular or diffuse or a combination of the two. The existence of different propagation time delays in the constituent wave reaching the receiver will result in independence in the statistical properties for signal frequencies sufficiently separated. If frequency components of the wideband spread spectrum signal experience such a statistial decorrelation, coherent addition of energy in the received waveform is no longer possible. If one of the stations in the link is moving, the received radio signal will in addition to multipath exhibit variations in amplitude due to changing multipath contributions and apparent frequency.

For some radio links with a predominant direct path it may be possible to treat multipath components, both specular and diffuse, as noise in the calculation of the signal/noise ratios. However, for scatter channels all of the energy reaches the receiver after some form of reflection and the only signal component available will experience fading. Under the assumption of slow fading, i.e. when coherence time is much greater than the data integration period and for which the signal bandwidth is less than the coherent bandwidth, the fading signal will be nonselective. In DS spread spectrum systems, multipath returns produce correlation responses which overlap the desired autocorrelation output. These responses and the effect they produce will depend on the delay time. For delays that are shorter than the spreading code chip duration the multipath may destructively interfere with the desired signal and cause both time varying and frequency selective fading. Multipath returns with delays exceeding the code chip duration may result in multiple autocorrelation outputs which may upset the synchronization and signal tracking procedures. If the delays are comparable with the data bit duration, intersymbol interference may occur if the spreading codes are not changed from data bit to data bit.

The effects of other user interference in code multiplexed systems will, as shown in Chapter 3, strongly depend on the structure of the codes employed. However, where fading is present, well designed spreading codes may not be sufficient to avoid incorrect data polarity decisions even in the absence of Gaussian noise.

This chapter will derive expressions for the signal and interference components useful in *SNR* calculations under the condition of both nonselective and frequency or time selective channels. For nonselective channels the capacity of a multiple user system with code multiplexing is simulated for Rician fading statistics. As an example of the effect of time dispersion on spread spectrum modulation, an experiment with wideband signals on an HF skywave channel is analysed. Finally the problem of receiver synchronization and tracking for time dispersive channels is discussed.

5.2 Characterization of the channel

The communication channel may be specified by a scattering function $\sigma(r, f)$, which describes the distribution of reflection cross-section in time delay and in Doppler shift. For example an HF channel can be envisaged as clusters of scatterers representing the different multipaths. Each individual path is composed of scatterers spread over a range delay and a Doppler interval. Fig. 5.1 illustrates such a channel scattering function.

The received energy is the superposition of the contributions from all the scatterers affecting the channel. In addition to the scattering function the channel can be characterized by a channel correlation function $R(t, s, \tau, \zeta)$, where t, s form a pair of time variables and τ, ζ are delay variables. For wide sense stationary channels the correlation functions are invariant under a translation in time, i.e. they depend on the variables t and s only through the difference $t - s$. For the uncorrelated scattering channel the scattering amplitudes of two different scatterers with different delays are uncorrelated. Hence, for a wide sense stationary uncorrelated scatter channel (WSSUS channel) [1],

$$R(t, s, \tau, \zeta) = Q(\tau, t - s)\delta(\tau - \zeta) \qquad (5.1)$$

where Q is the delay cross power spectral density. Taking the Fourier transform of Q gives

$$\int_{-\infty}^{\infty} Q \exp\left[-\mathrm{j}2\pi\tau(t - s)\right] \mathrm{d}(t - s) = \sigma(\tau, f) \tag{5.2}$$

which is the same as the channel scattering function with the range delay replaced by the corresponding time delay τ.

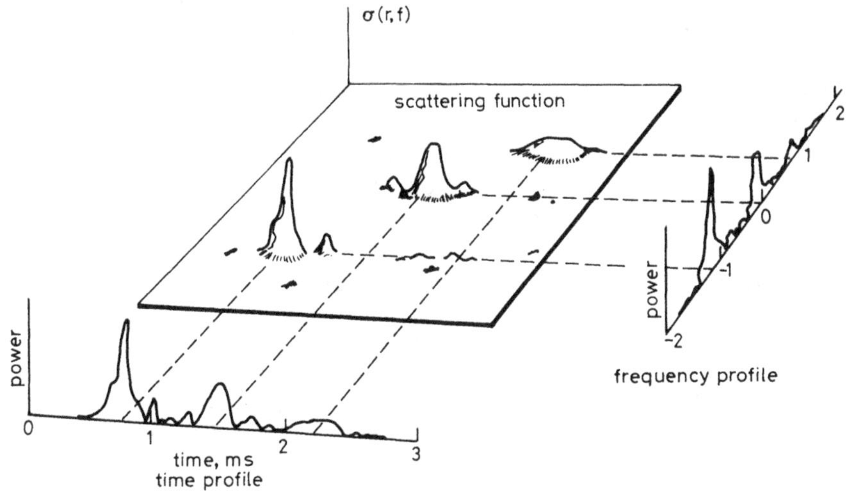

Fig. 5.1 *Time–frequency plane of the channel scattering function*

For channels with Doppler spread only (time selective) we can define a Doppler scattering function $S_\mathrm{D}(f)$ and a correlation function $Q(0, t - s)\delta(\tau)$ [2]:

$$Q(t - s) = \int_{-\infty}^{\infty} S_\mathrm{D}(f) \exp\left[\mathrm{j}2\pi f(t - s)\right] \mathrm{d}f \tag{5.3}$$

and

$$Q(0) = \sigma_\mathrm{x}^2 = \int_{-\infty}^{\infty} S_\mathrm{D}(f)\, \mathrm{d}f \tag{5.4}$$

where σ_x^2 represents the mean squared value of the underlying random process of the channel. For channels with only multipath we can define a range scatter function $S_\mathrm{r}(\lambda)$, where λ is increment in distance, which expresses the expected value of energy returned from an increment located at a certain distance.

Total received energy can then be expressed as

$$E_\mathrm{received} = E_\mathrm{transmitted} \int_{-\infty}^{\infty} S_\mathrm{r}(\lambda)\, \mathrm{d}\lambda \tag{5.5}$$

$$= E_\mathrm{transmitted}\sigma_\mathrm{x}^2 \tag{5.6}$$

Taking the Fourier transform of $S_\mathrm{r}(\lambda)$ we obtain the two-frequency correlation function

$$R_v = \int_{-\infty}^{\infty} S_\mathrm{r}(\lambda) \exp\left(-\mathrm{j}2\pi\lambda v\right) \mathrm{d}\lambda \tag{5.7}$$

The nondispersive channel may then finally be characterized by $\sigma(r, f) = \delta(r)\delta(f)$.

Consider a transmitted waveform $s(t)$ with duration T and bandwidth W. The waveform may be subjected to range delays of L seconds to Doppler D Hz. If $L \ll T$ there is no apparent time spreading of the received waveform. However, the signal may still be distorted if $TW \gg 1$ and $W \gg 1/L$. This is because it is possible for a waveform of bandwidth W to change its value significantly in a time $L \gg 1/W$. This means that it is possible that returns from some multipaths will interfere. Fig. 5.2 shows a segment from a PSK modulated spread spectrum signal, before and after transfer across a channel where $W^{-1} \ll L \ll T$. Thus it is possible for a spread spectrum system to degrade even without any apparent time dispersion. If the Doppler spread B is much less than T^{-1}, the signal waveform reciprocal duration, there is no apparent frequency spreading of the received waveform. If $TW \gg 1$ then $B \ll W$ will result in no dispersion, but if $BT \gg 1$ the received signal will be distorted.

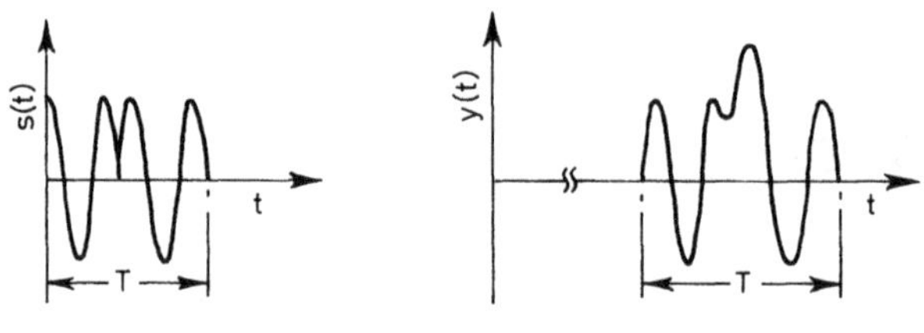

Fig. 5.2 *Waveform for channels for which $W^{-1} \ll L \ll T$. s(t) is the transmitted waveform, y(t) is the received waveform*

5.3 Spread spectrum system equations

5.3.1 *Doubly dispersive channels*

Consider a PSK modulated spread spectrum signal given by

$$s(t) = \sqrt{(2P)}a_1(t)b(t)\cos\omega_c t \tag{5.8}$$

where P is the transmitted power and ω_c is the carrier frequency. The spreading code is given by $a_1(t)$ and the data by $b(t)$. The data signal is defined by

$$b_i(t) = \sum_{n=-\infty}^{\infty} b_{in} p_T(t - nT) \tag{5.9}$$

where the information b_{in} are assumed to be independent identically distributed random variables taking values ± 1 with equal probability. p_T is the rectangular pulse function defined by

$$p_T(\lambda) = \begin{matrix} 1 & 0 < \lambda < T \\ 0 & \text{otherwise} \end{matrix} \tag{5.10}$$

The received signal after propagating over a fading channel is [3]

$$r(t) = \sqrt{(2P)}\, f(\tau, t - s) a_1(t - \tau) b_1(t - \tau) \cos(\omega_c t + \theta) + n(t) \tag{5.11}$$

where $f(\tau, t - s)$ represents the fading process introduced by the channel.

Detection will then depend on the realization of the matched filter. For SAW implementation detection is accomplished by multiplying a band limited version of the received signal with a local oscillator modulated by the time spreading code and then integrating over the SAW substrate length. For digital implementations it is usually required to lowpass filter the received signal, owing to speed limits in the digital circuitry, and then use inphase and quadrature techniques on a sampled and quantized version of the input signal. In the following the nonlinearities involved in the quantization are neglected and satisfactory sampling is assumed, so that the system equations used are similar to the familiar equations for active correlation at $t = T_d$.

The output then becomes

$$Y = \sqrt{(P/2)} \int_0^{T_d} a_1^2(t) b_1(t) f(\tau, t - s)\, dt + \int_0^{T_d} a_1(t) n(t) \cos \omega_c t\, dt \tag{5.12}$$

The second term is the noise and, since $n(t)$ is assumed to be a zero-mean Gaussian process, N will also have zero mean and a variance

$$\text{var}(N) = \frac{N_0}{2} \frac{T_d}{2}$$

where $N_0/2$ is the two-sided noise power spectral density at the input.

The SNR can now be defined in terms of probabilistic averages conditioned upon $b_i(t) = +1$ for $0 \leqslant t < T$.

For channels with no distinguishable direct signal component, the signal/noise ratio can be defined as

$$SNR = \frac{E(S^2)}{\text{var } N} = \frac{\text{var } S + [E(S)]^2}{\text{var } N} \tag{5.13}$$

where S represents the first term on the right-hand side of eqn 5.12.

Assuming the fading process to be at least short term stationary, then

$$E(S|b) = \sqrt{(P/2)}\, T_d E\{f(t)\} \tag{5.14}$$

where $E\{f(t)\}$ is the mean value of the fading envelope, which for convenience will be normalized to 1 in the following:

$$\text{var } S = P/2 \int_0^{T_d} \int_0^{T_d} \int_{-\infty}^{\infty} f(\tau, t - s) a_1(t - \tau) b_1(t - \tau) \times a_1(t) a_1(s - \tau) b_1(s - \tau) a_1(s)\, d\tau\, dt\, ds \tag{5.15}$$

If a multiple user system based on code division multiplexing is considered, the other user interference will enter into the denominator of the *SNR* expression, which now is given by

$$SNR_1 = \frac{E(S^2)}{\text{var } N + \text{var } I_1} \tag{5.16}$$

where var I_1 is given by

$$\begin{aligned} \text{var } I_1 = \frac{P}{2} \sum_{k=2}^{K} \Bigg[E \int_0^{T_d} \int_0^{T_d} \int_{-\infty}^{\infty} & f(\tau, t - s) a_k(t - \tau - \tau_k) \\ & \times b_k(t - \tau - \tau_k)\, a_1(t) a_k(s - \tau - \tau_k) \\ & \times b_k(s - \tau - \tau_k) a_1(s)\, d\tau\, dt\, ds \Bigg] \end{aligned} \tag{5.17}$$

If the channel considered contains a nonfading direct path and a specular and a diffuse multipath, the *SNR* will be modified to read

$$SNR_1 = \frac{E(S_d)}{(\text{var } N + \text{var } S_f + \text{var } I_1)^{1/2}} \tag{5.18}$$

where S_d now represents the direct path signal and S_f defines the fading portion of the signal. Var I_1 will now contain a fading and a nonfading component.

5.3.2 *Time selective channels*

If no time dispersion is present then $\tau = 0$, and for a signal subject to fading the expression for var S in eqn 5.15 is simplified to

$$\text{var } S = P/2 \int_0^{T_d} \int_0^{T_d} f(0, t - s) b(t) b(s)\, dt\, ds \tag{5.19}$$

Since it is the fading correlation as a function of $t - s$ which is of interest, one may replace $(t - s)$ by Δt. Then

$$\begin{aligned} \text{var } (S|b) &= P \int_0^{T_d} f(0, \Delta t)(T_d - \Delta t)\, d\Delta t \\ &= PT_d \int_0^{T_d} f(0, \Delta t) \left(1 - \frac{\Delta t}{T_d}\right) d\Delta t \\ &= PT_d \sigma_f^2 \int_0^{T_d} \left(1 - \frac{\Delta t}{T_d}\right) \frac{P(\Delta t)}{\sigma_f^2}\, d\Delta t \end{aligned} \tag{5.20}$$

where $P(\Delta t)/\sigma_f^2 = \varrho_f(\Delta t)$ is the normalized fading envelope autocovariance function and σ_f^2 is the fading envelope variance.

$$\varrho_f(\Delta t) = \frac{[f(t) - E(f)][f(t + \Delta t) - E(f)]}{\sigma_f^2} = \frac{R_f(\Delta t) - [E(f)^2]}{\sigma_f^2} \tag{5.21}$$

where $R_f(\Delta t)$ is the autocorrelation function for the envelope of f.

For the Gaussian process,

$$p_f^2(\Delta t) = \left(\frac{1}{\sigma_f^2}\right)^2 \{[x_I(t)x_I(t+\Delta t)]^2 + [x_Q(t)x_I(t+\Delta t)]^2\}$$

Let the envelope of the fading process be described by a Rayleigh distribution $x(t) = x_I(t)\cos\omega_0 t + x_Q(t)\sin\omega_0 t$. Then the inphase x_I and quadrature x_Q components are Gaussian distributed: $\overline{x_I(t)x_Q(t+\Delta t)} = 0$ for the narrowband case since x_I and x_Q are slowly varying compared with $\cos\omega_0 t$ and $\sin\omega_0 t$.

The autocorrelation function for the envelope of f can now be expressed in terms of a hypergeometric function

$$R_f(\Delta t) = F[-1/2, -1/2; 1; \varrho_x^2(\Delta t)] \tag{5.22}$$

where the hypergeometric function can be expanded in an infinite series. Ignoring terms beyond second order gives

$$R_f(\Delta t) = [1 + \tfrac{1}{4}\varrho_x^2(\Delta t)] \tag{5.23}$$

Choosing a Gaussian shaped autocorrelation function for the inphase and quadrature fading components gives

$$\varrho_x^2(\Delta t) = \exp\left(\frac{-\Delta t^2}{T^2}\right)$$

where T is the correlation time.

The signal/noise ratio is a function of the fading conditions. If the correlation time T is large compared with the data bit time T_d the fading envelope will be constant during the integration time and the result is no time selectivity. If T is short compared with T_d the signal statistics will approach a Gaussian distribution and the channel will be time selective.

In a multiple user system where code division multiplexing is used, the SNR is given by eqn 5.16:

$$SNR = \frac{E(S^2)}{\text{var}\, I + \text{var}\, N}$$

For the case of no time dispersion the variance of the interference in eqn 5.17 now becomes

$$\text{var}\, I_1 = \frac{P}{2}\sum_{k=2}^{K} E\left[\int_0^{T_d}\int_0^{T_d} f(0, t-s)a_k(t-\tau_k)b_k(t-\tau_k)a_1(t) \times a_k(s-\tau_k)b_k(s-\tau_k)a_1(s)\,dt\,ds\right] \tag{5.24}$$

The expectation of the data sequence $E\{b_k(t-\tau_k)b_k(s-\tau_k)\}$ will be different from zero only if $(t-\tau_k)$ and $(s-\tau_k)$ corresponds to the same data bit. This will only occur when $t > \tau_k$ and $s > \tau_k$ or when $t < \tau_k$ and $s < \tau_k$. Thus the inner integral will only contribute for part of the integration range $[0, T_d]$. It is therefore not surprising that for systems where the data bit duration is equal to the spreading code sequence period, eqn 5.24 reduces to [3]

$$\operatorname{var} I_1 = \frac{P}{2T_d} \sum_{k=2}^{K} \int_0^{T_d} \int_0^{T_d} f(0, t - s) C^k[T_d - (t - s)] a_1(t) a_1(s) \, dt \, ds \tag{5.25}$$

Changing to the variable $\Delta t = t - s$ gives

$$\operatorname{var} I_1 = \frac{P}{T} \sum_{k=2}^{K} \int_0^{T_d} f(0, \Delta t) C^k(T_d - \Delta t) C^i(T_d - \Delta t) \, d\Delta t \tag{5.26}$$

The interference variance and thus the final *SNR* will now depend on the fading correlation time as well as the structure of the spreading codes. For well designed codes the aperiodic value is small compared with the autocorrelation peak. Thus if the fading correlation time is longer than the code chip duration, the major contribution to the interference variance will occur for $\Delta t = \{0, T_c\}$. If the correlation time increases further the value of the interference variance will not change much.

If however the correlation time is small compared with the code chip duration, an averaging effect over the chip duration will take place and the interference variance will decrease.

5.3.3 *User capacities for code multiplexed systems with nonselective fading channels*

In Chapter 6 the system performance of CDMA networks is discussed; however, to illustrate the effects of nonselective fading on spread spectrum system capacity, a situation with slow fading is considered here. The correlation time is considered to be long compared with the data bit time, so in this case the fading will be constant over the range of integration. The model used [4] allows for Rician fading statistics specified by the division of power between a specular component and a Rayleigh fading component. Ratios of 10%, 50%, 70% and 90% are simulated. The distributions are shown in Fig. 5.3. They are generated by an analytic approximation to the Q function based on a Taylor's series expansion. The effective received powers P_i are the result of the transmitters matching their output power to the expected path loss including the average fading depths. Thus on the average all stations are received with the same power. The network is judged a success when all links in the network meet certain link criterions. A signal/interference threshold value T_h can be defined such that for ratios less than T_h the link is said to fail, whereas for ratios greater than T_h the link operates normally. With the received signal power S defined, it is useful to define the excess E in decibels such that $E = S - T_h$. This provides a link performance criterion such that if $E > 0$ the link will operate and if $E < 0$ it will fail.

Specific configurations can then be analysed according to this network performance criterion. Since the fading properties and code interference are of a statistical nature it is possible to randomly generate a set of configurations with the same properties. To provide a statistical estimate of the network success a particular number of stations in a network configuration is said to have a certain percentage of time of successful operation, defined as the time that number of stations is received with a specified *SNR* (threshold) at the receiver:

$$T^i_{\text{success}} = \frac{\sum_{u=1}^{e} W^i_u}{c}$$

$$W = 1 \quad \text{for } \frac{S^i}{N+I} \geqslant T_{\text{h}} \qquad (5.27)$$

$$= 0 \quad \text{otherwise}$$

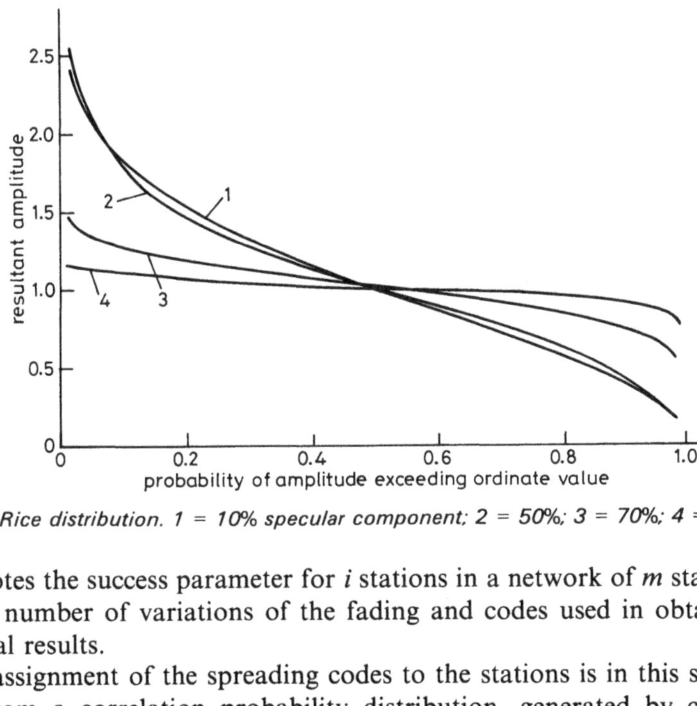

Fig. 5.3 *Rice distribution. 1 = 10% specular component; 2 = 50%; 3 = 70%; 4 = 90%*

W^i denotes the success parameter for i stations in a network of m stations and c is the number of variations of the fading and codes used in obtaining the statistical results.

The assignment of the spreading codes to the stations is in this simulation made from a correlation probability distribution, generated by computing correlation values for all relative phase shifts for 1023 bit Gold codes. For the simulation result shown in Figs 5.4 and 5.5, c is equal to 1000 and the number of stations in a network is 10, 20 and 40.

The results show that the fading gives a network design with possible severe near–far problems. For situations with small specular/scattered power ratios, the fading will greatly reduce the user capacities of a CDMA network.

5.3.4 *Frequency selective channels*

For frequency selective channels memory is exhibited and intersymbol interference is introduced. Limiting this effect to the two adjacent bits and assuming no frequency dispersion gives for the signal variance in eqn 5.15

$$\text{var } S = P/2 \int_{-T_d}^{T_d} f(\tau, 0) \int_0^{T_d} \int_0^{T_d} a_1(t-\tau)b_1(t-\tau) \times a_1(t)a_1(s-\tau)b_1(s-\tau)a_1(s)\, dt\, ds\, d\tau \quad (5.28)$$

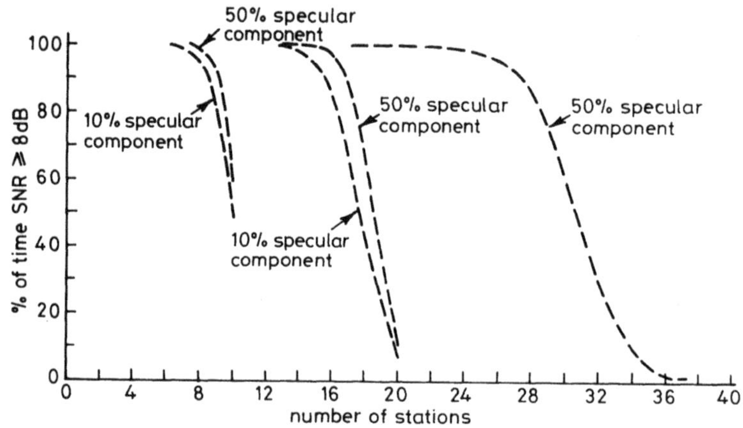

Fig. 5.4 *Performance of CDMA for 10, 20 and 40 stations and specular component = 10% and 50%. Required SNR = 8 dB*
(Reprinted from 'Operation of portable HF CDMA networks with controlled transmitter powers', *IEE Proceedings F*, vol. 127, no. 6, pp. 467–474, December, 1982.)

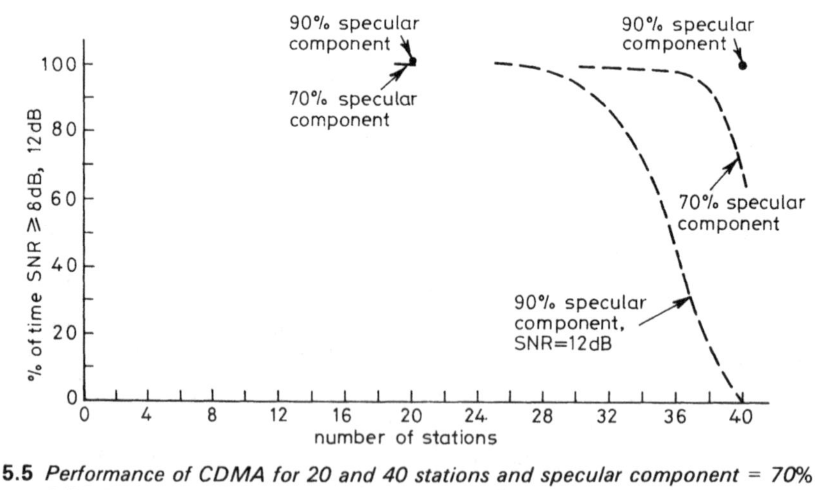

Fig. 5.5 *Performance of CDMA for 20 and 40 stations and specular component = 70% and 90%. Required SNR = 8 dB and 12 dB*
(Reprinted from 'Operation of portable HF CDMA networks with controlled transmitter powers', *IEE Proceedings F*, vol. 127, no. 6, pp. 467–474, December, 1982.)

Now the integral $\int_0^{T_d} a_1(t - \tau)b_1(t - \tau)a_1(t)\,dt$ is the periodic autocorrelation for the code a_1 if $b_1(t - \tau)$ is of the same sign for all $t \in [0, T_d]$. But from the assumptions made about the data bit stream in eqn 5.9, it will change sign with probability 0.5 for $t \in [0, T_d]$. Thus

$$\int_0^{T_d} a_1(t - \tau)b_1(t - \tau)a_1(t) = \begin{matrix} b_{1,0}C^1(T_d - \tau) + b_{1,-1}C^1(\tau), & \tau > 0 \\ b_{1,0}C^1(T_d + \tau) + b_{1,1}C^1(-\tau), & \tau < 0 \end{matrix} \tag{5.29}$$

Similar expressions can be given for the integral being a function of s. Then finally, averaging over data bit values of ± 1 assumed equally likely gives

$$\text{var } S = P/2 \int_0^{T_d} f(\tau, 0)[(C)^2(\tau) + (C)^2(T_d - \tau)]\,d\tau \tag{5.30}$$

Thus for frequency selective channels the *SNR* will directly depend on the aperiodic correlation properties of the spreading codes not only for multiple user systems but also for a single transmitter–receiver link.

In a multiplex system the variance of the interference in eqn 5.17 becomes

$$\text{var } I_1 = \frac{P}{2}\sum_{k=2}^{K} E\left[\int_0^{T_d}\int_0^{T_d}\int_{-\infty}^{\infty} f(\tau, 0)a_k(t - \tau - \tau_k)b_k(t - \tau - \tau_k)a_1(t) \times a_k(s - \tau - \tau_k)b_k(s - \tau - \tau_k)a_1(s)\,d\tau\,dt\,ds\right] \tag{5.31}$$

Now the integral $\int_0^{T_d} a_k(t - \tau_k - \tau)b_k(t - \tau_k - \tau)a_1(t)$ is the periodic cross-correlation for codes a_k and a_1 if $b_k(t - \tau_k - \tau)$ is the same for all $t \in [0, T]$.

Arguments similar to those for the signal variance over a frequency selective channel with respect to change of data bit polarity finally give

$$\text{var } I_1 = \frac{P}{2T_d}\sum_{k=2}^{K}\int_{-\infty}^{\infty} f(\tau, 0)\,d\tau \int_0^{T_d} [(C^{k1})^2(\tau) + (C^{k1})^2(T - \tau)]\,d\tau \tag{5.32}$$

Typically the HF skywave radio path is regarded as being time dispersive due to multipaths. In this situation the rate at which digital signalling can take place is usually restricted to make $L \ll T_d$ in order to reduce the waveform distortion to an acceptable level.

Although spread spectrum modulation makes it possible to resolve HF multipath structure via correlation processing, this may necessitate the use of large bandwidths to resolve small delays. However, there remains a dispersion owing to the fact that there is no unique reflection point in the ionsophere but rather a volume which is not uniformly ionized. The actual HF signal encompasses a spectrum of frequencies owing to the transmitted modulation. Because the phases velocity in an ionized medium depends on the electron density and the wave frequency, each spectral component experiences somewhat different phase perturbations and hence exhibits different amplitude and phase delays after propagating to the receiver. Merely the fact that the phase velocity and thus the refractive index depend upon the electron density and frequency of the wave means that the ionosphere is a dispersive medium. At some limit the system degrades because the received signal exhibits statistical decorrelation at different frequencies within the signal bandwidth. The ionospheric mode will thus limit the maximum 'useful' bandwidth available.

An example of an electron density model is shown in Fig. 5.6 where the group time delay varies with frequency because of dispersion. One way to describe the mode structure of the HF channel is to use oblique incidence ionograms.

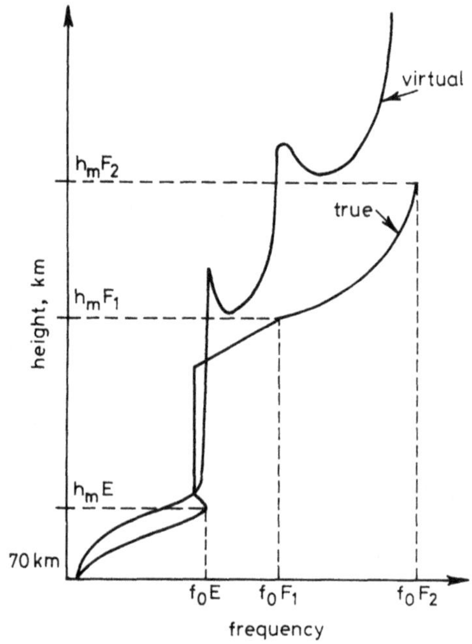

Fig. 5.6 *$N_e(h)$ model used in the IONCAP procedure*
(Courtesy of J. Lloyd, U. S. Department of Commerce, National Telecommunications and Information Administration.)

Fig. 5.7 shows an example of such an ionogram. Replacing the virtual height with a group time delay makes it possible to measure the dispersive nature of the ionosphere by calcualting the slope of the appropriate mode on such an ionogram.

Appendix D reports on experiments with spread spectrum modulation on an HF channel.

5.4 Synchronization and signal tracking in multipath environments

For the burst communication systems considered it has been assumed that the correlation detection and acquisition takes place in matched filters. For propagation environments with a direct path and static multipath channels, the problem of acquiring synchronization can be limited to the task of finding the largest correlation peak.

In this case one possible approach is given in Fig. 5.8. The two peak detectors are operated in parallel. During the search for the start of a message only one of the peak detectors is active. On the arrival of the first correlation peak

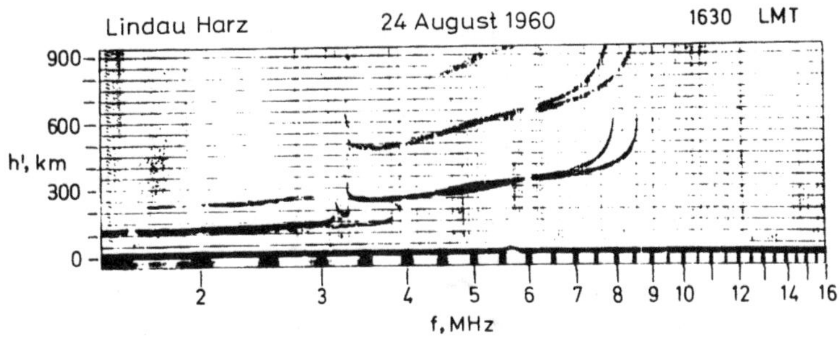

Fig. 5.7 *Quiet day ionogram*
(Courtesy of W. Becker, Max-Planck Institut für Ionospharen Physik, Lindau, West Germany.)

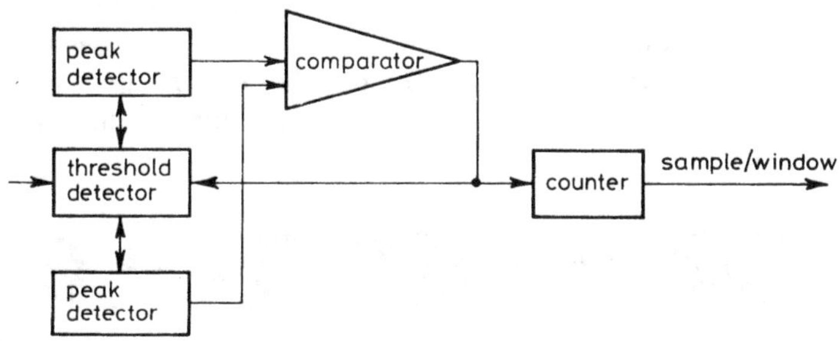

Fig. 5.8 *Sampling circuit for static multipath*

exceeding the threshold, its value is stored in the first peak detector. The exact point in time to be stored in the counters is determined by the threshold circuitry. On the negative going edge of the threshold device, the second peak detector is activated to register a possible new correlation peak within a time window equal to one data bit duration. If the comparator shows that the last peak is larger than the first, the positive going edge of the threshold detector will determine a new point of arrival into the counters. The correlation peak determined in this way can then be used as input to a data detector as described in Chapter 1. When the synchronization is completed, a window function is inserted and thereafter only the counter and threshold detector will be in action to adjust the sample point due to oscillator drift etc.

For channels without static multipaths, e.g. HF skywave links, a situation may occur where the effective propagation delay will change as different modes become dominant, causing synchronization problems. For short term fades of a given mode, the stability of the receiver clock could be made sufficient to allow automatic resynchronization when the fade is over. The data transmitted during the fade may however be lost owing to inadequate signal quality. However, permanent synchronization loss can occur with this scheme because the mode to which the receiver was initially locked has disappeared and other propagation modes have been established with different path delay times.

A possible system which accepts fadeout and is also able to search for new multipaths when the existing ones disappear has been tested on an HF skywave path. The synchronization and tracking system contains three main states. The synchronization circuit first searches for a correlation peak exceeding some threshold K_1. The value of K_1 is set to suit the required system SNR. When the threshold is exceeded the system will be at rest, and it is only the instability in the oscillators and random phase changes which alter the point of sampling. In the third state the correlation peak goes below a second threshold $K_2 < K_1$ defining an absolute minimum signal requirement. The system must now decide whether the cause of signal disappearance is a passing fading phenomena, in which case the system should remain in the second state after the interruption, or if a consistent phase error is the cause. If the latter is the case, the system should return to the first state. The decision of which state to go to can be based on signal time out. Fig. 5.9 shows the system in action for all these states. At point A the threshold is exceeded and the system tracks a particular multipath component. At some other point B this component starts to fade and at C the correlation peak goes below a second threshold. The system starts searching for a new multipath component which it detects at D, and the system starts tracking the new component.

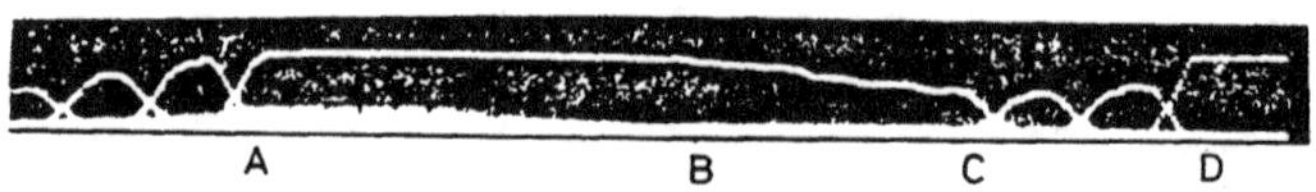

Fig. 5.9 *Demonstration of control unit*

For a condition of rapid changes in the multipath structure the system described would not work very well, since much time would be used to search for new and better modes or much data would be lost owing to inadequate signal quality.

The matched filter output is, in the case of resolved multipaths a series of pulses whose envelope is the approximate channel impulse response. If the characteristics of the ionosphere as defined by this impulse response are substantially unchanged for a data period, the energy from each of the propagation paths can be summed in a postdetection integrator. Such a modulation scheme achieves higher noise margin with more discrete multipath components. Fig. 5.10 shows a schematic receiver for DPSK modulation and Fig. 5.11 [5] shows the performance for the case of Rayleigh distributed signals. It is assumed that the interpath interference due to the spreading code aperiodic autocorrelation is negligible and that $\Delta \ll T_d$ so that no intersymbol interference occurs.

The postdetection integration concept can be refined to allow adaptivity, in the sense that a threshold device can determine the Δ integration period or the integrator may be weighted, recognizing that later paths are likely to contain less signal energy.

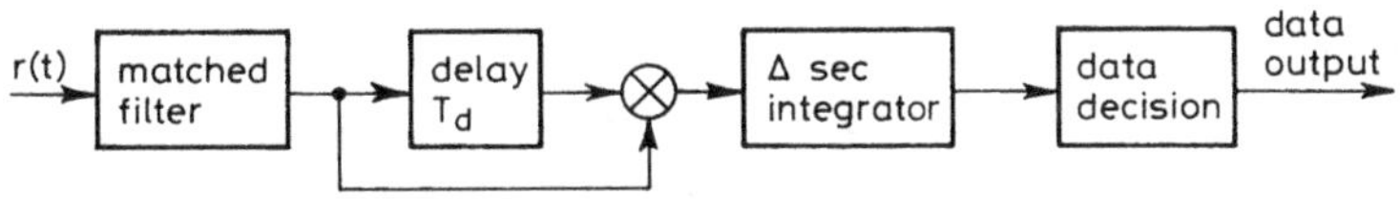

Fig. 5.10 *Post detection integrator for multipath channels*

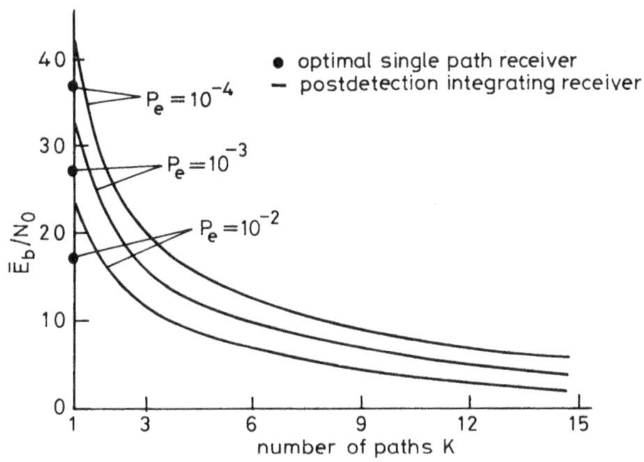

Fig. 5.11 *Required E_b/N_0 per path for DPSK with postdetection integrator in an environment of K independent Rayleigh fading paths. Noise power density N_0*
(TURIN, G. L., 'Introduction to spread spectrum anti-multipath techniques and their application for urban digital radio', *Proc. IEEE* 1980, **85** (3), pp. 328–353. Copyright © 1980 IEEE)

5.5 Filtering and equalization

Some of the channel effects introducing performance degradation in spread spectrum systems can be dealt with, to a large extent, through filtering or equalization. Adaptive equalizers are linear filter systems usually implemented with a tapped delay line structure. This provides a convenient adjustable parameter set which can be controlled to compensate for multipath interference. The equalization concept requires *a posteriori* statistics characteizing the channel transfer function. The statistics must in the noncoherent case provide estimates of path variables like relative delay and multipath strength. Usually the knowledge is obtained from separate sounding signals and circuitry or for the case of wideband spread spectrum modulation the data signals may in some cases be used themselves.

Under the resolvability condition, wideband spread spectrum signals can give accurate estimates of the multipath structure with respect to delay and amplitude. The heights of the correlation peaks are proportional to peak strengths and their relative time positions give the multipath profile.

A particular type of binary phase code, the 'convoluted' code briefly discussed in Chapter 3, permits the use of simple microprocessor based matched filters able to act as a sounding system with some low rate data communication capability.

Such a realization has been constructed [6] where the matched filter processing is carried out in a Z-80 single board computer. The transmitted signal is the convoluted code of 12 × 13 × 13 elements as shown in Fig. 5.12. The chip duration is 100 μs, giving an overall signal duration of ≈ 200 ms. The signal is transmitted about once per second and the received signal is I and Q sampled at a rate of 20 kHz and then stored in 16K of RAM. The stored signals are processed in the remaining ≈ 800 ms before the next transmission is received. To take care of multipath signals the input sampling is carried out for a certain period longer than the signal duration. The matched filter processing, taking place after all samples have been stored, is based on the tapped delay line configuration shown for two convoluted codes in Fig. 3.12. Since the storage elements are RAM the tap points are moved rather than shifting the signal from memory location to memory location.

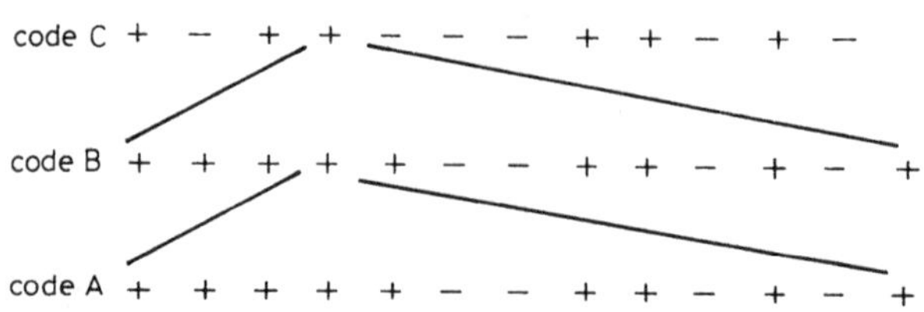

Fig. 5.12 *2028 bit convoluted code*

The processing commences after a block of data larger than the complete code length has been received and stored. Sums for all shifts of the outer code C are first computed and stored. These are then used in computing the sums for all shifts of code B, and these in turn are used to compute the final sums for code A. Since coefficients may have only the values + 1 or − 1, the coefficient multiplications reduce to additions and subtractions.

There are only 38 coefficients in total and, for speed, each is represented by its own inline code which adds to or subtracts from the accumulating sum.

The magnitude of the output is computed from the inphase and quadrature values. A digital/analogue converter provides an analogue signal for visual display on an oscilloscope.

The information carrying capability is achieved using a number of matched filters to determine which of a set of possible codes was sent.

The choice of which code, A, B or C, will represent the information depends on the autocorrelation function for the convoluted code. As Fig. 5.13 shows, the use of code A will make the inner sidelobe dependent on the information; the low sidelobe advantage, using for example Barker codes as inner codes, could be lost. The implemented system uses code C to convey information in order to retain the advantage of low sidelobes in sounding. The matched filter processing for each of the possible codes is carried out sequentially and the peak magnitudes are compared to determine which symbol was intended.

Unfortunately, although only code C is changed for each data symbol, the entire matched filter process must be repeated for each possible code since the code C process must occur first.

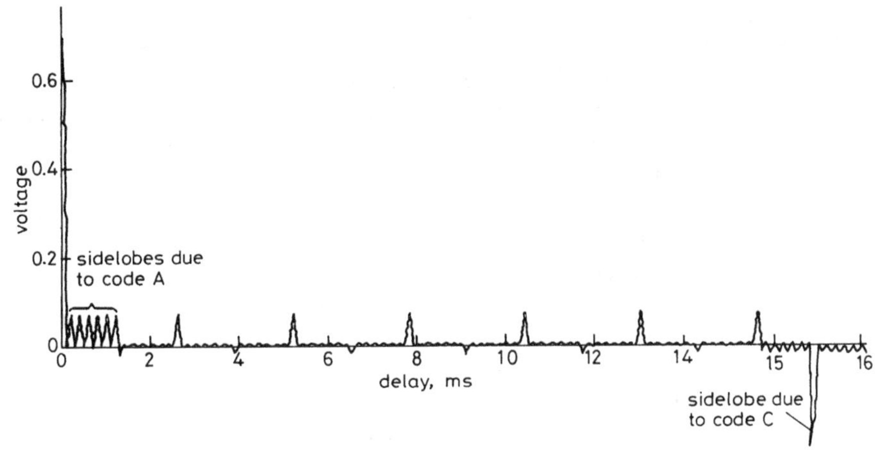

Fig. 5.13 *Autocorrelation function of 2028 bit convoluted code*

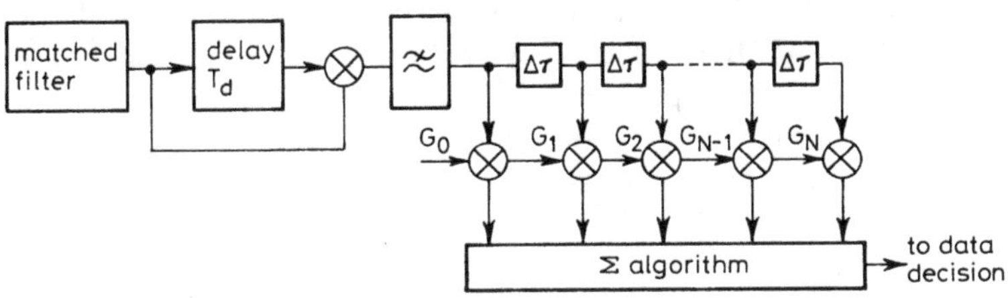

Fig. 5.14 *Matched filter and transversal filter as diversity combiner for resolvable multipath channels*

If all multipath delays $\tau < T_d$ for a spread spectrum system, then proper spreading code choice will give negligible intersymbol interference, and a transversal filter can be used after the lowpass filter as in Fig. 5.14. Before data transmission the spread spectrum system can transmit a sounding signal consisting of the unmodulated spreading code. The received signal will have multipath peaks as shown for example in Figs D.2, D.3 and D.4. This waveform will progress down the tapped delay line of the transversal filter. When the first correlation peak reaches the end of the delay line a threshold detector can initiate sampling of the waveform at each tap of the line exceeding some threshold. The path strengths can be read by peak detectors and these values and their position in the delay line can then be used to set the taps and gains to be used during data reception. The sounding procedure has to be repeated depending on the rate of change of the channel transfer function. Different combinations of the multipath samples are possible and usually reflect the expected multipath profile. The output of the transversal filter is thus a weighted sum of certain tap outputs given by the sounding cycle.

The length of the tapped delay line must be equal to the expected maximum delay τ seconds and should be tapped at least every $1/W$ seconds for a channel bandlimited to W, requiring a minimum of τW taps. The transversal filter has so far been acting as a diversity combiner and the spread spectrum modulator ensures adequate bandwidth to combat deep frequency selective fading. What happens if the path delays are no longer less than the data bit period? In this case intersymbol interference will occur even for a spread spectrum system; this will be reduced to the crosscorrelation between spreading codes if they are changed from data bit to data bit. However in many situations it is not possible to change the spreading codes in this way. The delay line has taps every $1/W$ seconds and for a spread spectrum system $T_d W \gg 1$. For intersymbol interference to occur $\tau > T_d$ so that $W\tau \gg 1$, which means that the number of taps will be large. For a spread spectrum system with a time–bandwidth product $T_d W = 1000$ and a multipath environment with $\tau_{max} = 2T_d$, the number of taps will be $(1000/T_d)2T_d = 2000$ taps. Obviously even in a 'heavy' multipath environment only a small percentage of those taps would be activated at any time, and it becomes unlikely that correlation peaks belonging to adjacent data bits will be aligned with an activated tap when a particular data bit is read out. Thus the transversal filter is able to suppress intersymbol interference. It should be noted that for the case $\tau > T_d$ the sounding cycle during which the taps are set must use a spreading code with a period long enough to ensure that its duration is longer than the maximum multipath delay.

The transversal filter and correlator configuration would be able to resolve and combine multipaths on HF skywave links that are due to multiple reflection layers and multihops etc. However, Fig. 5.3 shows that each path in itself is time dispersive. The slope of the ionogram will indicate the degree of dispersion. The dispersive nature of each path is not resolvable with a spread spectrum sounding system like the one previously described. Making use of bandwidths large enough to be affected by the group time delay slope will, as shown in Appendix D, result in loss of correlation peaks. To increase the spreading code chip rate above the coherent bandwidth requires an equalizer in front of the matched filter, as shown in Fig. 5.15. One configuration, usually easy to implement, is the inverse filter equalizer, which in fact can equally well be described as another transversal filter.

The filter must calculate

$$h(t) = r(t) * v(t) = s(t)$$

where $r(t)$ is the received signal, $v(t)$ is the appropriate weighting function and

$s(t)$ is the transmitted signal. During the sounding cycle the equalizer obtains samples of the bandlimited channel transfer function $c(t)$ after deconvolving the known sounding signal $s(t)$ and $c(t)$. The tap weights for the data receiving mode are then obtained through inversion of the channel transfer function. The sounding signal can either be time impulsing or frequency domain sampling through the use of Fourier transform techniques.

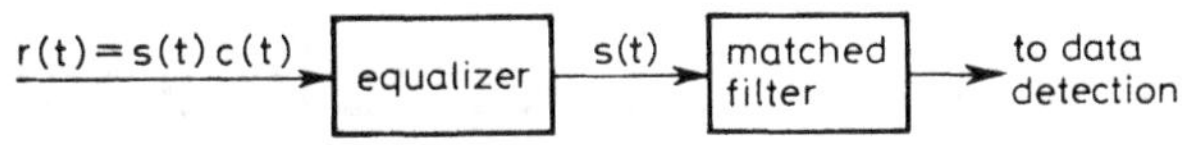

Fig. 5.15 *Equalized receiver*

Experiments [7] seem to indicate that the HF channel transfer function may be fairly constant for about 10 s and that bandwidths of 1 MHz are achievable with a single inverse filter equalizer.

References

1 BELLO, A. P.: 'Characterization of randomly time-variant linear channels', *IEEE Trans.*, 1963, **CS-11** (4), pp. 360–394

2 KENNEDY, R.S.: 'Fading dispersive communication channels' (Wiley Interscience, 1969)

3 GARDNER, C.H. and ORR, J. A.: 'Fading effects on the performance of a spread spectrum, multiple access, communication system', *IEEE Trans.*, 1979, **COM-27** (1), pp. 143–149

4 SKAUG, R.: 'User capacities of portable HF CDMA networks with controlled transmitter powers', Communication Research Centre, Report no. 1350, Ottawa, November 1981.

5 TURIN, G. L.: 'Introduction to spread spectrum antimultipath techniques and their application for urban digital radio', *Proc. IEEE*, 1980, **85** (3), pp. 328–353.

6 VENIER, G. O.: 'A simplified matched filter for large time–bandwidth sounders'. *IEEE Conf. Publ. 206*, 1982, pp. 136–140

7 DHAR, S. and PERRY, B. D.: 'Equalized megahertz-bandwidth HF-channels for spread spectrum communications'. Progress in Spread Spectrum Communications, IEEE Military Communications Conference, October 1982, Section 29.5–2.

Chapter 6

Code division multiple access networks

6.1 Introduction

It was shown in chapter 3 that the processing gain achieved using large spreading chip/data bit ratios could efficiently be utilized by transmitting many signals simultaneously on the same carrier frequency and applying code division multiplexing. Successful use of spread spectrum code division multiple access techniques requires the construction of spreading codes giving rise to a minimum of interlink interference. However, situations arise where the effects of interlink interference are amplified owing to operational considerations. Consider a network operated in a master–slave configuration. Let slave station M_1 transmit the desired signal S to the master, which is d_1 km away, and let another slave station M_2 at a distance d_2 km from the master transmit a signal I which is interference to the M_1–master link. Let C (dB) be the signal/interference ratio (S/I) required at the terminals of the master station's receiving input to produce the desired output S/I. The requirement is that

$$S - I \geqslant C \text{ (in dB)} \tag{6.1}$$

With transmitting powers of P_{M1} and P_{M2} and path losses of L_1 and L_2 respectively, eqn 6.1 after rearranging yields

$$(P_{M1} - L_1) - (P_{M2} - L_2) \geqslant C \tag{6.2a}$$

$$L_2 \geqslant C + (P_{M2} - P_{M1}) + L_1 \tag{6.2b}$$

Eqn 6.2*b* is an explicit formula for the near–far problem and will put a restriction on where slave station M_2 can operate if the original system specification is to be met. If the transmitter powers remain constant while the value of L_2 is reduced, the inequality in eqn 6.2*a* may be violated for any practical code design. The value chosen for the factor C will however depend on how well the codes can be designed. The near–far problem for multiple access networks is discussed in Section 6.2, and Section 6.3 evaluates the system performance for

CDMA. Section 6.4 is a note on the derivation of code subsets from Gold codes to increase user capacities.

The large bandwidths associated with spread spectrum systems make it unlikely that they will be the sole users present within a band. Narrowband communication systems are likely to interfere with the spread spectrum networks. The extreme sensitivity of propagation loss with respect to range for some groundwave or line-of-sight frequencies makes likely processing gain values unable to 'win' the power battle. It might therefore be imperative to find a system solution where narrowband interference is reduced before spread spectrum processing. Some available techniques to achieve this goal are discussed in Section 6.5.

6.2 System performance criteria for code division multiple access

In CDMA networks all members communicate on the same frequency. Usually there is also a common data rate $R_d = 1/T_d$ and a common spreading code chip rate $R_c = 1/T_c$. In burst transmission networks no overall timing reference to enable chip synchronization is usually achievable. In portable systems a probe signal from a possible time master would not be a sufficiently accurate time reference, since one would lack compensation for the different delays in the various transmission paths. Thus it is unlikely that the spreading codes would be chip synchronized.

The receiver will attempt to extract the individual spreading codes from a composite of many during the matched filter processing. It was shown in Chapter 3 that both the periodic (even) and odd crosscorrelation functions play key roles in calculating system performance for such situations. The wideband input signal consists of the wanted signal as well as the interfering signals, each spread by their own code. The spreading code modulating the wanted signal will match the receiver filter and the correlation peak will be sampled at a rate equal to the data rate. If the unwanted signals were totally uncorrelated with the wanted signal, then they would produce no correlation peaks at the filter output. However, the effect of crosscorrelation between the local sequence and the sequences of the unwanted signals appears as crosscorrelation peaks at the output of the filter. From Section 3.4 the signal/noise plus interference ratio in a CDMA system is given by

$$[S/(N + I)]_K = \frac{E[Y^k \mid b_0^k]^2}{\sigma_I^2 + \sigma_n^2} \tag{6.3}$$

where K is the total number of users and σ_I^2 and σ_n^2 are the variances of the other user interference and thermal noise respectively.

From eqn 6.3 it is clear that the actual SNR is degraded from that of a single link, where the only interference is the thermal and atmospheric noise.

Inserting in eqn 6.3 the variances from Section 3.4, i.e.

$$\sigma_n^2 = \frac{N_0 T_d}{4}$$

$$\sigma_I^2 = \frac{1}{24p} \sum_{\substack{r=1 \\ r \neq k}}^{K} (A^r)^2 S^{kr}$$

$$E[Y^k]^2 = \tfrac{1}{4}(A^k)^2 T_d^2$$

gives

$$[S/(N + I)]_K = \frac{\tfrac{1}{4}(A^k)^2 T_d^2}{\left[\dfrac{(A^k)^2}{24p} \displaystyle\sum_{\substack{r=1 \\ r \neq k}}^{K} \alpha^r S^{kr}\right] + \dfrac{N_0 T_d}{4}} \tag{6.4}$$

where $(A^r)^2 = \alpha(A^k)^2$, so that α is the power ratio between the rth interfering user and the desired signal.

An important question now is how much the signal power must be increased to maintain the signal/noise ratio for which the system is designed. The answer will obviously depend on the number of users, the code design and on α^r, the power ratio between the received signals. The new value for the thermal signal/noise ratio required to meet the original system specification is obtained from eqn 6.4 by rearranging

$$2E/N_0 = \frac{S/(N + I)}{1 - [S/(N + I)\dfrac{1}{6p^3} \displaystyle\sum_{\substack{r=1 \\ r \neq k}}^{K} \alpha^r S^{kr}} \tag{6.5}$$

E is the signal energy, defined as $\int_0^{T_d} s^2(t)\,dt = (A^2/2)\,T_d$, and $2E/N_0$ represents the thermal signal/noise ratio for a matched filter at the instant of sampling.

The increase in signal power required by each user to maintain the same performance in a multiple access system as it would without the power increase in a single-link system is called the *power cost of code multiplexing*. Each user in the CDMA network is forced to increase its output power by the amount disclosed by the power cost figure.

Assuming now for a moment that all interferring users are received with the same signal power $\alpha(A^k)^2$, then $\sum_{r=1, r \neq k}^{K} \alpha^r S^{kr}$ can be replaced by $(K - 1)\alpha \sum_{r=1, r \neq k}^{K} S^{kr}$. Solving now for K, the number of possible users, gives

$$K = \frac{1}{[S/(N + I)]\,\alpha\left(\dfrac{1}{6p^3}\right) \displaystyle\sum_{\substack{r=1 \\ r \neq k}}^{K} S^{kr}} - \frac{1}{(2E/N_0)\,\alpha\left(\dfrac{1}{6p^3}\right) \displaystyle\sum_{\substack{r=1 \\ r \neq k}}^{K} S^{kr}} + 1 \tag{6.6}$$

The factor $S/(N + I)$ is just the threshold value that is designed for and is fixed for a given performance. Thus the number of users will for a given threshold value depend on an increased thermal *SNR*, α, and the code design. If the other user interference is small compared with the thermal noise, then any increase in thermal *SNR* will increase the possible number of users for a given $S/(N + I)$ threshold. However, when the other user interference is dominant, the second term in eqn 6.6 approaches zero and the number of users is only determined by the required threshold, code design, and α. The latter condition is usually called the *bandwidth starved condition*.

In practical situations α will of course not be the same for all the interfering signals, and the solution of eqn 6.6 will be much more cumbersome and will generally require computer processing.

6.3 Evaluation of system performance

Section 6.2 developed the basic set of equations required to evaluate user capacities and performance for a CDMA network. In this section a network simulation model will be developed to quantify the near–far problem experienced in many CDMA applications [1]. Both the periodic (even) and odd crosscorrelation is part of the interference parameter S^{kr} in eqn 6.4. The odd correlation is present because asynchronous operation may result in a situation where the elements of the spreading code belonging to the required signal are correlated against the spreading code elements from interfering signals, which might be out of phase with the data bit edges of the required signal. This makes the crosscorrelation result sensitive to the presence of data transitions in the interfering signals. There are no analytical results which give the values of the odd crosscorrelation for common spreading codes. Some upper and lower bounds are given in Chapter 3 but they do not give a tight bound in general. Furthermore, there is a difficulty with these bounds because they do not take into account the phase sensitivity of the odd crosscorrelation, which means that phases can be selected to reduce the odd crosscorrelation. However, the periodic crosscorrelation of, for example, Gold codes is well described analytically by its crosscorrelation spectrum as shown in Chapter 3. In practice, burst formatted message signalling usually starts each message with a preamble consisting of all ones or all zeros data to synchronize. Since synchronization capacity is important, it is common to first select a general class of codes which have good periodic crosscorrelation, for example Gold codes, and then search in the hope of finding a subfamily for which the odd crosscorrelation is also small. Because of the difficulties in obtaining the odd correlation, and because the mathematical variance of Q^{kr} indicates a much larger interference fluctuation ratio and thus code dependence for cases with even crosscorrelation, the analysis carried out in this section restricts the code interference to periodic crosscorrelation values. However, performance degradation for CDMA due to other user interference based on both even and odd correlations is shown for different spreading sequences and no near–far effect in Section 3.6.6.

The interference will be the spectral components of the product of spreading codes influencing the sampling at the output of the matched filter, or equivalently spectral components falling inside a filter matched to the data rate.

The autocorrelations of the interfering signals are each given by:

$$R_I(\tau) = \frac{1}{T_c}(T_c - \tau) \qquad \tau < T_c \tag{6.7}$$

when the interference has the same centre frequency as the desired signal.

The power spectral density is then given by:

$$S(\omega) = 2\int_0^{\infty} R_I(\tau)\, e^{-i\omega\tau}\, d\tau \tag{6.8}$$

Taking the real part of $S(\omega)$ gives:

$$\begin{aligned} S(\omega) &= \frac{2}{T_c}\int_0^{T_c} (T_c - \tau)\cos\omega\tau\, d\tau \\ &= \frac{2}{\omega^2 T_c}[1 - \cos\omega T_c] \\ &= T_c \frac{\sin^2(\omega T_c/2)}{(\omega T_c/2)^2} \end{aligned} \tag{6.9}$$

If the ratio T_d/T_c is large, then the only noise falling inside a filter matched to the data rate will be in the vicinity of zero frequency (DC). Thus putting $\omega = 0$ gives

$$S(0) = T_c \tag{6.10}$$

Assuming that the power in the unwanted signal is P_i, the amount falling inside a filter matched to the data rate is given by

$$P_i' = P\frac{T_c}{T_d} \tag{16.11}$$

This is the same performance as with white Gaussian noise interference. In practice eqn 6.11 is only valid for codes with periods longer than the data bit duration and for large processing gains. When the code sequence period is equal to the data bit period the only spectral line inside the final filter is the DC component, its magnitude being represented by the periodic crosscorrelation value. If any Doppler or oscillator drift is present, additional Fourier components of the code product can be present at the output of the filter. Generally the RMS interference voltage is given by:

$$V_I = \left[\sum_K (\theta^{kr})^2 + mA^2\right]^{1/2} \qquad r \in [K] \tag{6.12}$$

where θ^{kr} is the periodic crosscorrelation for sequence k and sequences r in the set of K sequences, m is the number of spectral components of the codes $k + r$

inside the information bandwidth, and A is the amplitude of these spectral components. Now m is given by $(pT_c/T_d) - 1$ and the amplitude A will, close to zero frequency, approximate $\sqrt{(1/p)}$. Thus;

$$V_I = \left[\sum_K (\theta^{kr})^2 + \left(\frac{T_c}{T_d} - \frac{1}{p}\right)\right]^{1/2} \tag{6.13}$$

So far the treatment has assumed that the interferers are in perfect chip synchronism with the wanted code. Allowing for an arbitrary displacement by some fraction of a chip gives

$$V_I = \left\{\sum_K [\theta^{kr'} + (\theta^{kr''} - \theta^{kr'})t_c]^2 + (2t_c^2 - 2t_c + 1)\left(\frac{T_c}{T_d} - \frac{1}{p}\right)\right\}^{1/2} \tag{6.14}$$

where $\theta^{kr'}$ and $\theta^{kr''}$ is the crosscorrelation between two codes for two consecutive chip shifts, and t_c is a fraction of a chip.

Then when $T_c/T_d = 1/p$,

$$V_I \approx \left\{\sum_K [\theta^{kr'} + (\theta^{kr''} - \theta^{kr'})t_c]^2\right\}^{1/2}$$

or when Doppler or frequency offset is present,

$$V_I \approx \left\{\sum_K [\theta^{kr'} + (\theta^{kr''} - \theta^{kr'})t_c]^2 + (2t_c^2 - 2t_c + 1)\left(\frac{T_c}{T_d} - \frac{1}{p}\right)\right\}^{1/2} \tag{6.15}$$

Finally, when $T_c/T_d \gg 1/p$ and $T_c/T_d \ll 1$, then θ^{kr} will be small and

$$V_I \approx \left[\sum_K (2t_c^2 - 2t_c + 1)\left(\frac{T_c}{T_d} - \frac{1}{p}\right)\right]^{1/2} \tag{6.16}$$

which is equivalent to the white Gaussian noise case.

It is expected that, owing to the fact that only the periodic crosscorrelation value is used, a somewhat different number of possible CDMA users will result than might be realized in a practical system. However, this simplfication will not obscure the relative importance of the near–far problem to be modelled. For the purpose of this model interest is centred on a network configuration such as that shown in Fig. 6.1 where a number of stations at some time are in simultaneous communication with one station.

For such a situation there is usually assumed to be a sharp signal/interference threshold above which the probability of successful operation (for example specified error rate) tends rapidly to unity and below which the probability tends rapidly to zero. Because the transition from successful operation to failure occurs over a narrow range of signal/interference ratios, it is reasonable to define a threshold value T_h such that for ratios less than this value the link fails, whereas for ratios greater than this threshold the link operates normally. This simple approach can be used to define the successful and unsuccessful operation of each link so that

$$\begin{aligned} &[S^k/(N_0 + I)]_k \geqslant T_h \qquad k\text{th link successful} \\ &[S^k/(N_0 + I)]_k < T_h \qquad k\text{th link unsuccessful} \end{aligned} \tag{6.17}$$

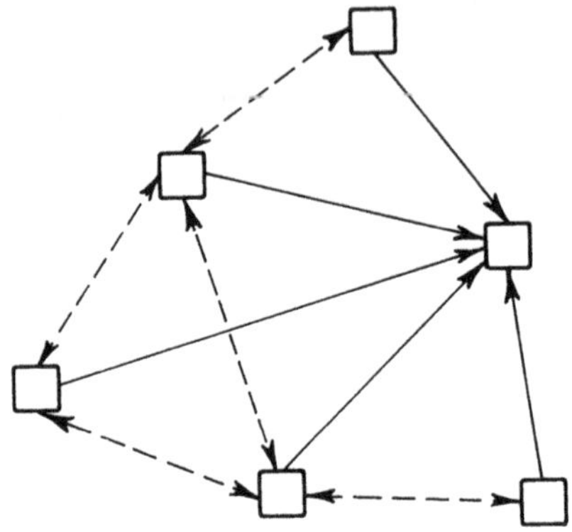

Fig. 6.1 *Network model for CDMA application*

The network may then be judged a success when all links in the network meet the link criterion. Using this definition the performance of the network is determined by its ability to support links or stations. This puts emphasis on the feature of CDMA networks that is of primary interest, namely the number of users to be supported.

6.3.1 *Simulation*

Specific locations and code assignments are randomly generated. The stations are assumed to be normally distributed and a configuration of the network can be obtained by specifying the number of stations, the mean and the standard deviation relative to the receiver of interest.

The assignment of the code sequence is made from the crosscorrelation probability distribution. In the results reported here, Gold codes with period 1023 were used with a correlation spectrum as in Fig. 6.2. The procedure has been to study each configuration for a number of statistical variations, thus providing a statistical estimate of the network success. A network configuration is said to have a probability p_c of successful operation if all the slaves are received with the signal/noise ratio required for p_c of 100 trials. The 100 trials picture different possible configurations with the same number of stations, mean path length and standard deviation of path length.

Thus

$$P_c = \sum_{u=1}^{100} \frac{A_u}{100} \qquad A = \begin{cases} 1 & \text{for } S^k/(N_0 + I) \geqslant T_h \text{ for all } k \\ 0 & \text{otherwise} \end{cases} \tag{6.18}$$

where T_h is the required threshold for link k.

The model allows a wide range of propagation frequencies, but the results disclosed here are based on a 10 MHz radio frequency assuming groundwave propagation. The parameterization of the propagation loss follows the CCIR 10 MHz propagation curve.

Fig. 6.2 *Correlation spectrum for Gold codes with period 1023*

It is assumed that the 10 MHz groundwave signal will not be subject to fading. The values used to represent atmospheric noise are fixed median values. The statistical distribution of the individual noise value is not considered since all results are relative to a single-link performance. Thus median noise values will be used to obtain typical powers and user capacities for CDMA. For a low speed 50 bit/s network, a reference temperature of 300 K and an atmospheric noise level of 44 dB above kT_0B results in a total noise power of -140 dBW.

6.3.2 *Discussion of results*

Three consecutive assumptions are made for the network transmitters during simulation. First, the transmitters will use their maximum and identical powers without consideration of their actual distance to the receiver. Secondly, the stations will make transmitter power adjustments to roughly reflect their distance to the receiver. Such an adjustment is assumed to result in a transmitter power within 10 dB above the minimum power necessary to match the propagation loss in question and the required *SNR*. Finally, the transmitters are able to more or less match their powers to the transmitter–receiver path length and the required *SNR*.

The results, with power cost, number of users, power matching accuracy and path length spread as parameters, are plotted on a graph system using 'mean distance from slaves to master' and 'percentage of success of network' as the x and y variables, respectively.

The plots in Figs 6.3 and 6.4 show the performance for 10 and 20 stations with a path length spread as defined by the normal distribution with a 10 km standard deviation. Curves are shown for different power matching accuracies and different values of the power cost (*PC*). The curves in Figs 6.5 and 6.6 show the result of another simulation run with 10 and 20 stations, but now the path length spread has been increased and is defined by a distribution with a standard deviation of 20 km. Finally, Fig. 6.7 shows the result for a network configuration of 50 stations and a path length spread defined by a distribution with a 10 km standard deviation.

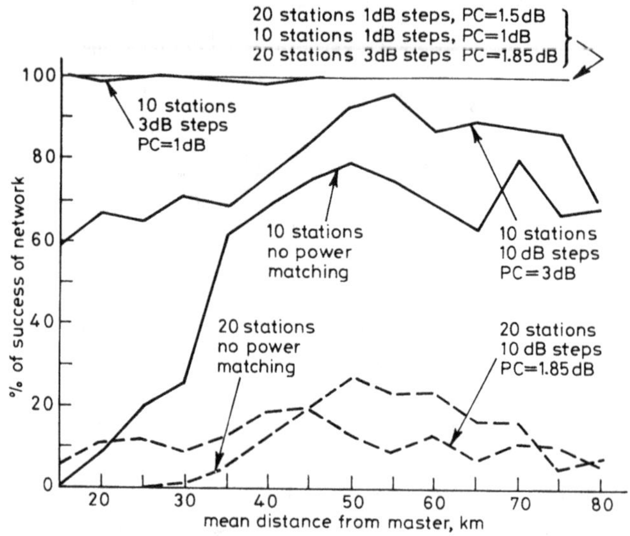

Fig. 6.3 *Simulation results showing degree of success of network with 10 and 20 stations*
(Reprinted from 'Operation of portable HF CDMA networks with controlled transmitter powers', *IEE Proceedings F*, vol. 127, no. 6, pp. 467–474, December, 1982.)

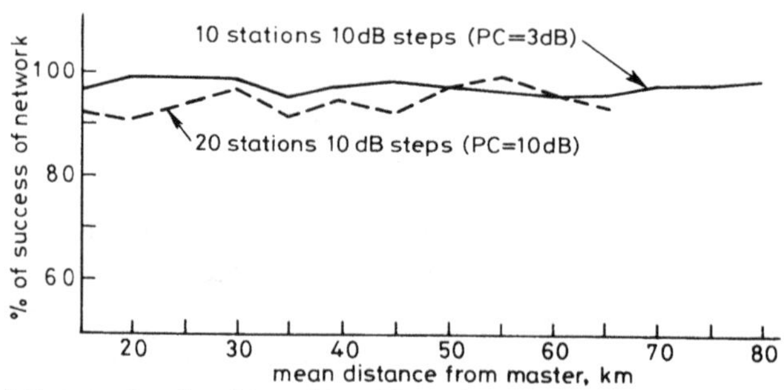

Fig. 6.4 *Simulation results showing achievable performance using coarse power control but large transmitter powers*
(Reprinted from 'Operation of portable HF CDMA networks with controlled transmitter powers', *IEE Proceedings F*, vol. 127, no. 6, pp. 467–474, December, 1982.)

The use of coarse power matching accuracy or no power matching at all does introduce significant performance degradation compared with good power matching accuracy and the same power cost. The penalty for obtaining good performance with coarse power matching accuracy is a large increase in transmitter powers within the network. The degradation is, for the case of 10 and 20 stations and the smaller geographic spread, most severe at the shorter mean distances. The reason for this is that, owing to the shape of the path loss curve, the differential loss will depend on the mean distance from the master. This

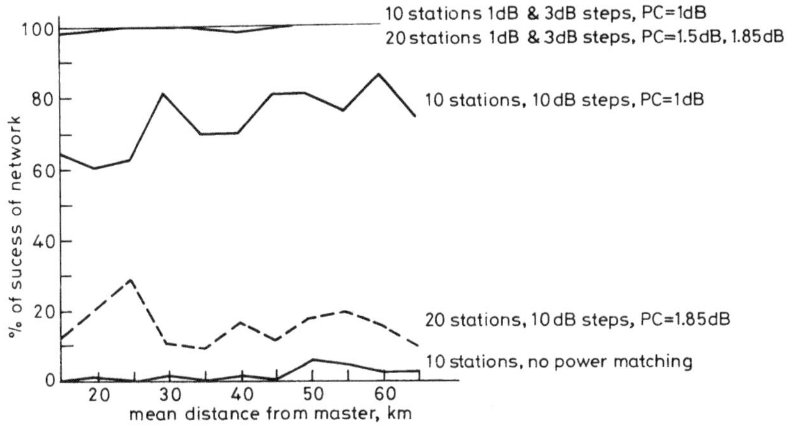

Fig. 6.5 *Simulation results showing degree of success of network with 10 and 20 stations being widely separated*
(Reprinted from 'Operation of portable HF CDMA networks with controlled transmitter powers', *IEE Proceedings F*, vol. 127, no. 6, pp. 467–474, December, 1982.)

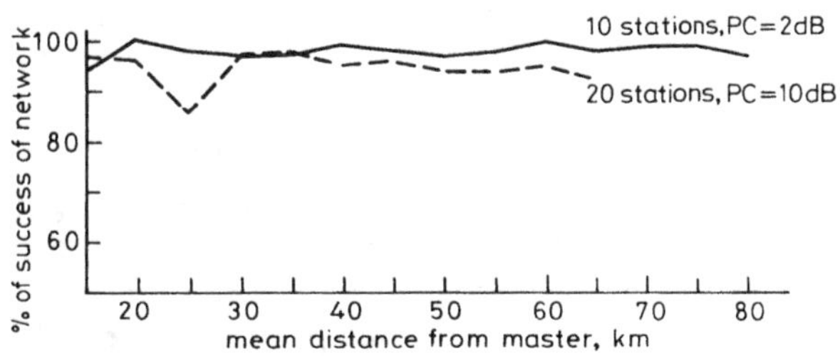

Fig. 6.6 *Simulation results showing achievable performance using 10 dB power step sizes but large transmitter powers*
(Reprinted from 'Operation of portable HF CDMA networks with controlled transmitter powers', *IEE Proceedings F*, vol. 127, no. 6, pp. 467–474, December, 1982.)

makes the need for power matching less important at longer mean distances. For the larger geographic spread the effect of the distance dependent incremental path loss becomes less distinct owing to a larger range of distances for a given mean. For the case of a network with 50 stations it becomes impossible to operate with much success using coarse power matching accuracy, whatever power cost is used.

With 50 users the total other user interference is comparable with the noise power and the channel is bandwidth starved. The path losses encountered in groundwave propagation are very sensitive to the path distances and will as the simulations show for most practical cases introduce near–far problems. The maximum possible number of users will depend mainly on the accuracy of the power control system. However, it may generally be concluded that for situations where the other user interference is less than the radio system's noise level, poor power control can be compensated for by increased transmitter output powers. With no power control at all, the CDMA system deteriorates quickly as the number of users increases.

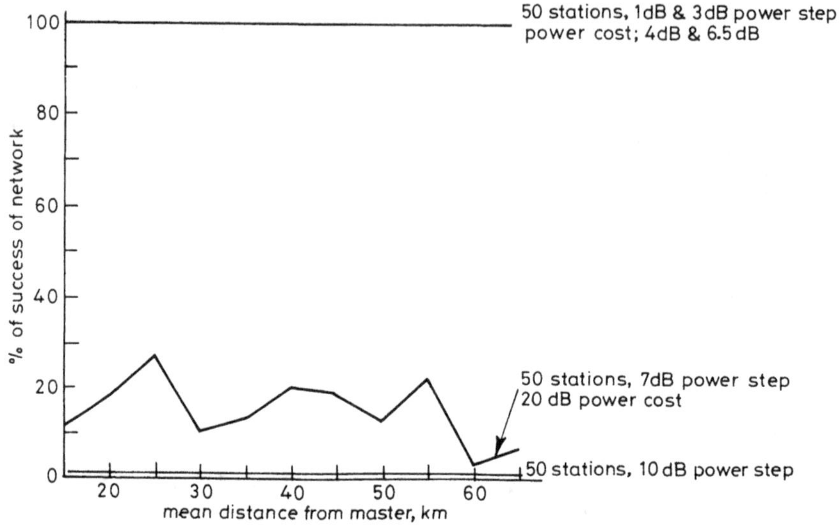

Fig. 6.7 *Simulation results showing degree of success of network with 50 stations* (Reprinted from 'Operation of portable HF CDMA networks with controlled transmitter powers', *IEE Proceedings F*, vol. 127, no. 6, pp. 467–474, December, 1982.)

A question to ask is how well in practice a transmitter–receiver pair is able, if required, to match the output powers to the actual path length. Some simple on-the-air experiments have been conducted to answer this. The trials took place at HF groundwave frequencies and distances. The equipment consisted of two HF transceivers with programmable output powers connected to two microprocessors. The measurement equipment formed a closed loop. One of the stations transmits a call signal. The second station receives this signal and, from a given algorithm, calculates the power to get back to the first station with a predetermined strength. The call signal was FSK modulated in the test arrangement, and was envelope detected and sampled for A/D conversion. A variable number of signal and noise samples could then be collected and averaged in a signal/noise ratio calculation. The value of necessary power to meet system requirements is then calculated in the microprocessor and is used to program the output power. Either the noise levels can be assumed equal at the two sites, or information on the noise level can be transmitted as part of the call signal.

The difference between the system specified S/N and the S/N achieved with the power matching algorithm was collected as an error output. The experiment was repeated a number of times at 5–20 MHz to collect statistics on the error.

The simple experiment showed that power matching could be achieved with a departure from specified S/N values indicated by a standard deviation of approximately 2.5 dB for 32 signal/noise samples and reduced to approximately 1 dB for 256 samples.

If it is assumed that the signal levels are constant and the noise is Gaussian distributed, the error in the output power will be less than one standard deviation for approximately 70% of the time.

6.4 Optimization of code design for CDMA

The simulation results of Section 6.3.2 considered Gold codes for multiplexing, and the other user interference is calculated from periodic crosscorrelation values. Chapter 3 discusses the merits of Gold codes related to their periodic crosscorrelation spectra. In this section the characterization will be extended to include average crosscorrelation values, and it is shown that this leads to a possible sieve for Gold codes in CDMA applications.

The nonmaximal codes described by Gold do have a bound on their peak periodic crosscorrelation, but the average value of θ^{kr} (averaging taking place across all relative phase shifts) can vary considerably among code pairs from the same family. Such an average value is a possible measure of interference in a network, as discussed in Section 6.3. In a burst communication system the individual codes will, over time, collide under all relative phase shifts.

The imbalance of 1s and 0s in the Gold codes is bounded by t since the imbalance is not more than the correlation of the generating m-sequences. Thus the imbalance takes on the values -1 and $-1 \pm 2^{(m+e)/2}$. Thus

$$\sum_{l=0}^{p-1} \theta^{G_iG_j}(l) =$$

$$\begin{aligned}
&= +1 && \text{if } p - 2\text{wt}(G_i) = p - 2\text{wt}(G_j) = -1 \\
&= 1 - 2^{(m+e)/2} && \text{if } p - 2\text{wt}(G_i) = -1,\ p - 2\text{wt}(G_j) = -1 + 2^{(m+e)/2} \\
&= 1 + 2^{(m+e)/2} && \text{if } p - 2\text{wt}(G_i) = -1,\ p - 2\text{wt}(G_j) = -1 - 2^{(m+e)/2} \\
&= 1 + 2^{(m+e)} - 2^{(m+e+2)/2} && \text{if } [p - 2\text{wt}(G_i)] = [p - 2\text{wt}(G_j)] = -1 + 2^{(m+e)/2} \\
&= +1 + 2^{(m+e)} + 2^{(m+e+2)/2} && \text{if } p - 2\text{wt}(G_i) = p - 2\text{wt}(G_j) = -1 - 2^{(m+e)/2} \\
&= 1 - 2^{(m+e)} && \text{if } p - 2\text{wt}(G_i) = -1 + 2^{(m+e)/2} \\
& && \quad\ \ p - 2\text{wt}(G_j) = -1 - 2^{(m+e)/2}
\end{aligned}$$

where G_i and G_j denote the Gold codes i and j respectively, and wt(G) denotes the Hamming weight of code G.

Hence, depending on the Gold codes chosen, the average value of $\theta^{G_iG_j}$ can change by a factor of $|1 - 2^{(m+e)}|$.

Because of this wide spread it may be worth while to pick a subset of nonmaxial sequences generated from Gold's theorem, to avoid the large possible average crosscorrelation. The following example illustrates this.

A preferred pair of m-sequences with periods 1023 is to be used. Let α be the root of $x^{10} + x^7 + 1$. Then $t = 2^{(10+2)/2} + 1 = 65$ and α^{65} is a root of $x^{10} +$

$x^8 + x^7 + x^4 + x^2 + x + 1$. The distribution of crosscorrelation values between this preferred pair of m-sequences is as follows:

value -1 occurs $2^{10} - 2^{10-2} - 1 = 767$ times

value $-1 + 2^{(10+2)/2} = 63$ occurs $2^{10-2-1} + 2^{(10-2-2)/2} = 136$ times

value $-1 - 2^{(10+2)/2} = -65$ occurs $2^{10-2-1} - 2^{(10-2-2)/2} = 120$ times

Thus the crosscorrelation averaged over all shifts equals:

$$(-1) \times 767 + 63 \times 136 + (-65) \times 120 = 1$$

Table 6.1 *Distribution of imbalances of the cross-correlation of nonmaximal Gold codes*

Number of −65	63	−65	63
151	103	118	133
120	135	112	127
120	135	124	139
122	137	155	107
120	135	118	133
114	129	132	147
118	133	147	99
124	139	116	131
122	137	110	125
124	139	120	135
126	141	120	135
114	129	92	170
94	172	165	117
159	111	142	157
128	143	133	85
149	101	116	131
82	160	143	95
94	172	145	97
130	145	114	129
112	127	90	168
128	143	120	135
120	135	120	135
124	139	157	109
118	133	90	168
124	139	126	141

Then the first nonmaxial sequence generated by the preferred pair of m-sequences is correlated with all the other (1022) nonmaxial sequences for all possible shifts. Table 6.1 shows the number of times the values −65 and 63 occur for the first 50 nonmaxial sequences. As can be seen by inspection, they all depart from the correlation value distribution for the preferred m-sequence pair.

The crosscorrelation averaged over all shifts takes on only three values, -63, 3969 and -4095. The imbalances in the nonmaxial sequences will vary between -1, 63 and -65. The crosscorrelation averaged over all shifts is equal to the product of the imbalance in the selected sequences.

For all the nonmaxial sequences the pairwise possible average crosscorrelation values are

1 $(-1) \times (-1) = 1$
2 $63 \times (-1) = -63$
3 $(-65) \times (-1) = 65$
4 $63 \times 63 = 3969$
5 $(-65) \times (63) = -4095$
6 $(-65) \times (-65) = 4225$

It is possible to select balanced Gold codes (the number of 1s exceeding the number of 0s by 1) and hence minimize average crosscorrelation. First the preferred pair of m-sequences must be placed in their characteristic phases, identified by the property that in this phase they are invariant under decimation by 2^h, $h = 0, 1, 2, \ldots, (n - 1)$. The relative phases of the m-sequence pair necessary to generate balanced Gold codes are then given by any relative shifts, so that the initial 1 in the first sequence corresponds to a 0 in the other sequence.

The previous discussion is based on the assumption that the signs of the crosscorrelation values are retained through the processing. However, in many practical applications coherent demodulation is not feasible and inphase and quadrature techniques with signal squaring are used, removing the sign of the crosscorrelation functions. In this case averaging of crosscorrelation magnitudes would be a useful measure of performance. The mean square value of the crosscorrelation is defined as

$$M = \sum_{l=0}^{p-1} |\theta^{ab}(l)|^2 = \theta^a(0)\theta^b(0) + \sum_{l=1}^{p-1} \theta^a(l)\theta^b(l) \tag{6.19}$$

which equals $p^2 + p - 1$ for maximal length sequences because

$$\begin{aligned} \theta^a = \theta^b &= p \qquad \text{if } l \equiv 0 \bmod p \\ &= -1 \qquad \text{if } l \not\equiv 0 \bmod p \end{aligned}$$

The mean square value for nonmaxial Gold codes can depart considerably from this value. It is possible to isolate sequences having a lower mean square value than those of preferred m-sequence pairs as well as those having a larger value.

Let $(a_i)_0^{p-1}$ denote the sequence $a_0, a_1, \ldots, a_{p-1}$ and $d(a_i)$ and $\mathrm{wt}(a_i)$ denote the Hamming distance and weights functions respectively. Then

$$\begin{aligned} \theta^{G_1G_2}(\tau) &= p - 2d[(a_{i+l} + b_i)_0^{p-1}, (a_{i+\tau} + b_{i+\tau})_0^{p-1}] \\ &= p - 2\mathrm{wt}[(a_{i+l} + b_i + a_{i+\tau} + b_{i+\tau})_0^{p-1}] \end{aligned} \tag{6.20}$$

Thus

$$\theta^{G_1G_2}(l) \quad = \quad -1 \qquad \text{and} \qquad \theta^{G_1G_2}(0) \quad = \quad -1$$

For other values of τ,

$$(a_{i+l} + a_{i+\tau})_0^{p-1} \quad = \quad (a_{i+k})_0^{p-1}$$

for some k depending on τ, owing to the add and shift property of m-sequences, and

$$(b_i + b_{i+\tau})_0^{p-1} \quad = \quad (b_{i+m})_0^{p-1}$$

for some m depending on τ. However, k cannot take the value l and m cannot take the value 0.

The factor $k(\tau) - m(\tau)$ will take on specific values in $[0, p - 1]$. The values will depend on l as well as the specific m-sequence pair involved. The sequence pair $G_1 G_2$ will then have a three-valued or in some cases two-valued crosscorrelation function. Thus a careful selection of l may lead to sequence subsets with a lower crosscorrelation value than the bound given by Gold. However, of greater concern here is the cardinality of the set

$$\{\tau : |\theta^{G_1G_2}(\tau)| \quad = \quad -1 \pm 2^{\lfloor(n+2)/2\rfloor}\} \tag{6.21}$$

The factor $k(\tau) - m(\tau)$ may be the same for several τ depending on the l and the m-sequence pair involved. Thus $\theta^{G_1G_2}(\ )$ may not take on the same set of values the same number of times as $\theta^{m_1m_2}(\ )$.

Certainly $\theta^{m_1m_2}(l)$ and $\theta^{m_1m_2}(0)$ will be replaced by -1 where $\theta^{m_1m_2}(l) = p - 2\text{wt}\,[(a_{i+l} + b_i)_0^{p-1}]$ and $\theta^{m_1m_2}(0) = p - 2\text{wt}\,[(a_i + b_i)_0^{p-1}]$.

These facts explain why the mean square value of the crosscorrelation between some Gold codes may depart significantly from the mean square value for pair of m-sequences.

A special case for Gold codes occurs when one of the sequences is a member of the preferred m-sequence pair, say a_i, and the other is generated by $a_{i+l} + b_i$. Then the previous calculation gives

$$\sum_{\tau=0}^{p-1} [\theta^{G_1G_2}(\tau)]^2 \quad = \quad \sum_{\tau=0}^{p-1} [\theta^{m_1m_2}(\tau)]^2 [\theta^{m_1m_2}(l)]^2$$

This makes it possible to generate a set of Gold sequences where the mean square value is reduced by t^2 compared with m-sequence pairs.

However, no easy way is known to upper or lower bound the cardinality in the general case.

The expected value of θ_{ab} is equal to zero for random binary sequences but $E\{\theta_{ab}^2\} = p$ and $\text{var}\,\{\theta_{ab}^2\} \approx 2p^2$. Then $E\{M\} = \sum_{l=0}^{p-1} E[(\theta^{ab})^2] = p$ and $\text{var}\,[(\theta^{ab})^2] \approx 2p^2$. For Gold sequences of period 63 a variation of the mean square value observed is given by $E\{M\} - 0.3\sigma_M < M < E\{M\} + 0.5\sigma_M$.

For sequences of period 1023 the variation observed is given by $E\{M\} - 0.16\sigma_M < M < E\{M\} + 0.14\sigma_M$.

Table 6.2 shows the mean square crosscorrelation for selected sequences with periods from 31 to 1023. In order to see the effect of using subsets with different average crosscorrelation, 'good' and 'bad' families were selected and then used in the simulation described in Section 6.3.1. Fig. 6.8 shows one result.

Table 6.2 *Mean crosscorrelation for selected Gold codes*

N	p	min D	max D	Mean square
5	31	10	23	735–1455
6	63	11	24	2783–6079
7	127	55	75	13919–19039
9	511	244	287	228863–293311
10	1023	223	283	912511–1158271

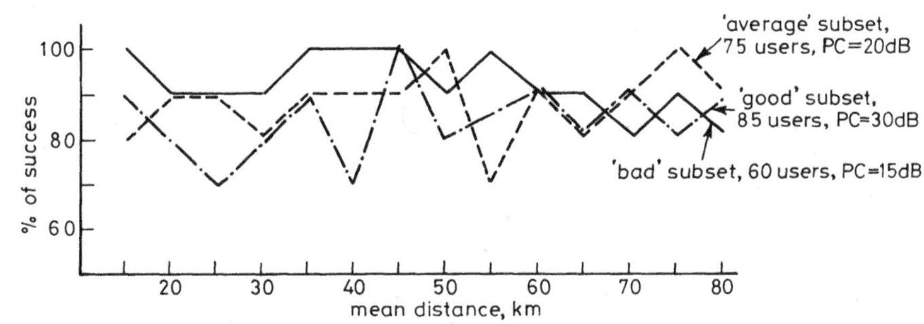

Fig. 6.8 *Performance of CDMA for selected spreading Gold codes, period 1023 bit. Required SNR = 11 dB, power matching = 0.01 dB. Results for 10 iterations. Best possible performance curves: increasing PC does not improve performance*

6.5 Narrowband interference suppression

It would obviously be very advantageous to system performance if one could reduce the level of narrowband interference (occasional from conventional radios or deliberate from jammers) before matched filter processing. This is of course just as applicable to single links as to CDMA networks. A possible solution is a notch filter placed at the centre frequency of the interference. Since one does not know in advance where the interference frequency is, the notch system must automatically be tuned to the right frequency.

The notch filters will necessarily degrade the correlation peak. Let $s(t)$ be the wanted spread spectrum signal which is passed through a filter with impulse response $h(t)$. The output of the filter becomes

$$y(t) = \int_0^{\infty} h(t)\, s(t - \tau)\, d\tau \tag{6.22}$$

This signal is then fed to the matched filter performing the correlation:

$$Y(t) = \int_0^{T_d} \int_0^{\infty} h(\tau)\, s(t - \tau) s(t)\, d\tau\, dt \qquad (6.23)$$

But

$$\int_0^{T_d} s(t - \tau) s(t)\, dt = R_T(\tau) \qquad (6.24)$$

which is the autocorrelation of the spreading sequence. Thus

$$Y(T) = \int_0^{\infty} h(\tau) R_T(\tau)\, d\tau \qquad (6.25)$$

and the filter impulse response will affect the output correlation value.

Experiments with fourth-order Butterworth filters have been reported [2] where the notch is placed either at the centre of the spreading bandwidth or with the centre displaced by one half the chip rate. Fig. 6.9 shows the degradation in decibels of the correlation peak as a function of the 10 dB notch bandwidth. Section 6.5.1 describes the result from an experimental automatic notch filter system based on phase locked loop techniques.

Another interesting approach, more able to cope with multi-interferers, is transformation of the total time signal to a representation in frequency where interferences placed at different frequencies are 'spread out' in time. A limiter can now in its simplest form be used to reduce interference 'peaks' to a preset threshold. Better, but possibly more difficult to achieve, the limiter can be made dynamic with a threshold which for positive S/N ratios follows the envelope of the useful signal, or for normal operation with negative S/N ratios follows the background noise level. The setting of the threshold should be fast enough to follow the signal envelope or noise level but should not be able to adjust to narrowband interference peaks.

Alternatively with added hardware complexity the 'time representation' of the frequency components can be used to gate out the unwanted interference. It should be possible to obtain time gates with very fast rise and fall times. This is similar to notch filters with extremely sharp filter characteristics. It is presently possible to realize the necessary time–frequency transformation using chirp z Fourier transformation implemented in SAW technology. For an input signal $s(t)$ the Fourier transform can be implemented in the following way. First the signal is multiplied with a linear chirp described by $\exp(-j\mu t^2)$, where $2\mu/2\pi$ defines the chirp rate in hertz per second. The resulting signal $s(t) \exp(-j\mu t^2)$ is then convolved with another chirp $\exp(j\mu t^2)$ in the form of a SAW filter with $\exp(j\mu t^2)$ as its impulse response. The output after convolution becomes

$$\int_{-\infty}^{\infty} s(\tau) \exp(-j\mu\tau^2) \exp[j\mu(t - \tau)^2]\, d\tau \qquad (6.26)$$

$$= \exp(j\mu t^2) \int_{-\infty}^{\infty} s(\tau) \exp(-j2\mu t\tau)\, d\tau \qquad (6.27)$$

where $2\mu t/2\pi$ has dimensions (Hz/s)s = Hz and represents the constant frequency in a time–frequency Fourier transform. The result after convolution is thus

$$\exp(j\mu t^2)\, S(2\mu t) \tag{6.28}$$

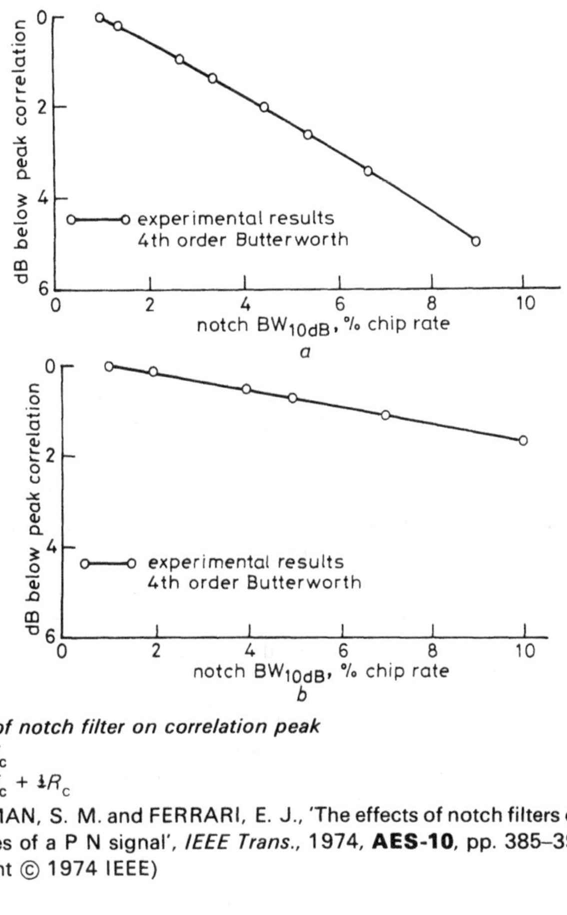

Fig. 6.9 *Effects of notch filter on correlation peak*
a $f_0 = f_c$
b $f_0 = f_c + \frac{1}{4}R_c$
(SUSSMAN, S. M. and FERRARI, E. J., 'The effects of notch filters on the correlation properties of a P N signal', *IEEE Trans.*, 1974, **AES-10**, pp. 385–390. Copyright © 1974 IEEE)

Then finally multiplication of this signal with a chirp of the form $\exp(-j\mu t^2)$ gives the required Fourier transform $S(2\mu t)$. Fig. 6.10 illustrates the transform circuitry.

At this stage the limiting or time gating function is introduced to reduce the effects of the narrowband interference. Then a similar but inverse transformation takes place before the result is fed to the matched filter. Fig. 6.11 gives a time–bandwidth representation for the signal passing through the complete transformation system. One should at this point note certain inherent limitations in such a system. The input signal will have a finite bandwidth which may be analysed for a finite time in the chirp filter. The longer the time of analysis is, the better is the frequency resolution. A problem, however, is that the time is finite, and this will have the same effect as a window function in more familiar Fourier analysis. It is, however, possible to alter the shape of the window function during the design of the SAW chirp line. A straightforward square windowfunction will introduce a broadening and a sin x/x shape of the narrow-band interference, and the sidelobes will be down only by some 13 dB from the mainlobe, as shown in Fig. 6.12. These high sidelobes would put severe constraints on the dynamic range of the interference for which the transformation system would give an overall improvement.

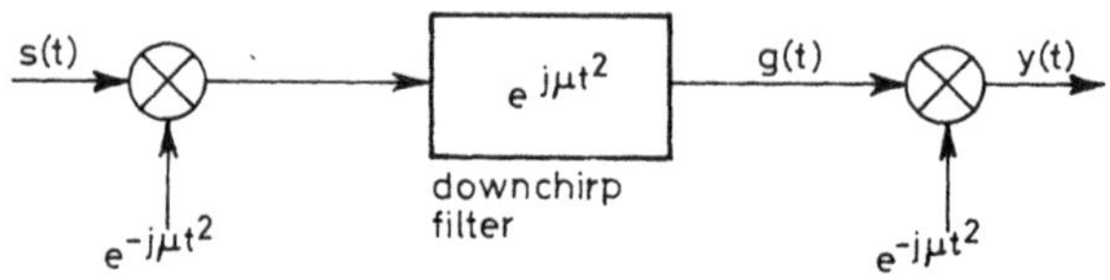

Fig. 6.10 *Transform circuitry using frequency sweeps and convolution*

It is therefore important to choose an optimal window function. If the limiter threshold is made to follow the wanted signal envelope or background noise, this requirement is less important. A system for interference suppression using SAW technology is described in Section 6.5.2.

A third possibility for reducing the effect of interference is to make use of spatial filtering in the form of steerable null antennas. Generally antenna array processing falls into two categories. The open loop system relies on the setting of predetermined weights which force the array output into a known form. Closed loop signal processing, however, derives an error signal from the actual final array output which is used during the adaptive processing. Fig. 6.13 shows an open loop system in its simplest form with two omnidirectional antennas able to eliminate one interference source. In addition to the antennas the circuitry consists of a phase adjusting network, a network to adjust the amplitude and a summation network. If the interference source and the direction to the wanted signal are sufficiently separated it will be possible to adjust the amplitude and phase of the network in such a way that the interference is suppressed relative to the wanted signal.

The relation between the necessary phase adjustment and the required position of the antenna null follows from an expression where the sum of the network phase displacement and the phase difference related to the direction of the interference source and the antenna separation should be 180°:

$$\phi + (2\pi d/\lambda)\cos\delta = \pi \tag{6.29}$$

where d is the antenna separation, δ is the angle formed by the direction to the interference source and the antenna base line, λ is the signal wavelength and ϕ is the necessary phase adjustment in the network.

The required value of ϕ as a function of δ is shown in Fig. 6.14 with d as a parameter.

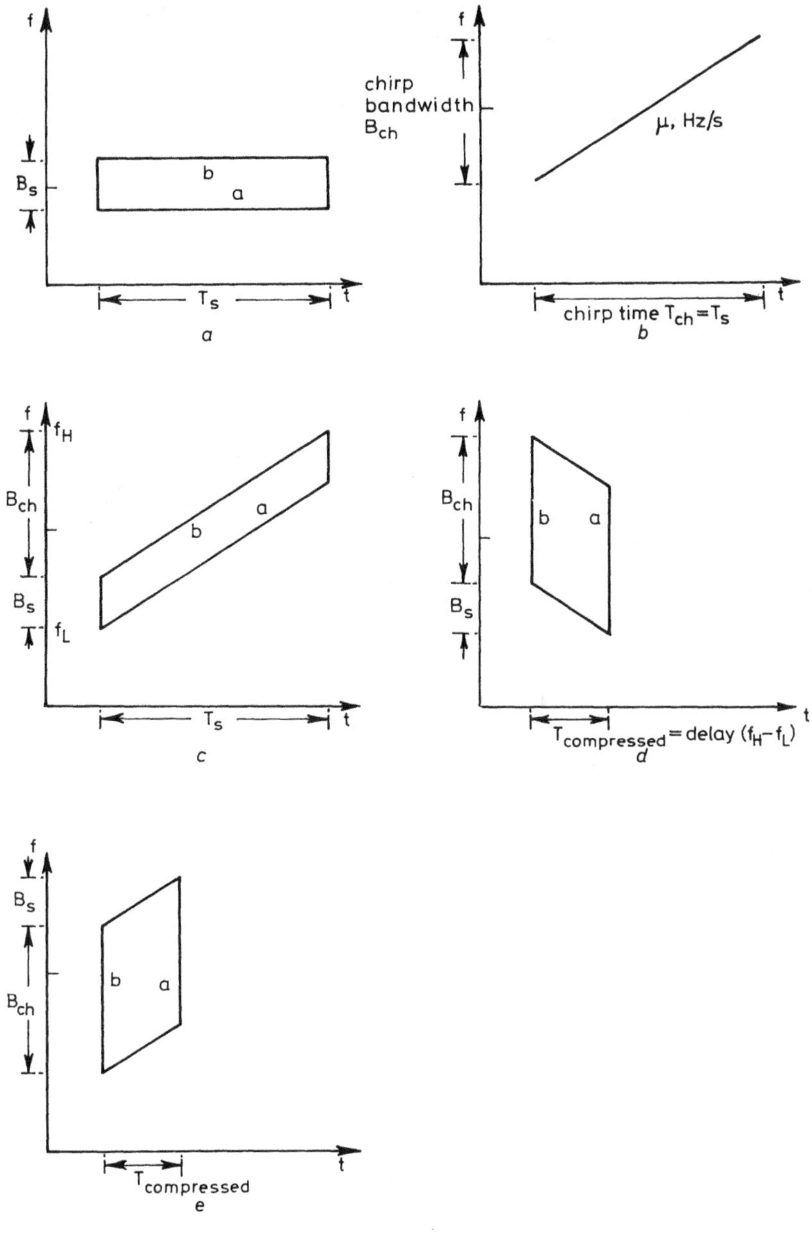

Fig. 6.11 *Time–bandwidth representation for transform circuitry*
a Input signal $s(t)$
b Linear chirp $\exp(-j\mu t^2)$
c $s(t)\exp(-j\mu t^2)$
d $S(2\mu t)\exp(j\mu t^2)$
e $S(2\mu t)$

The resulting antenna diagram obtained by summing the electric field vectors in the two signal branches is given by

$$\vec{E} = \vec{E}_1 + \vec{E}_2 \sim 2\vec{E}_1 \cos\left(\frac{\phi + 2\pi d \cos\delta/\lambda}{2}\right) \tag{6.30}$$

when $E_1 \sim E_2$ and ϕ is adjusted to put a null in the direction given by δ. If ψ now is any direction relative to the direction of the interference, the antenna power diagram is given by

$$E^2 = 4E_1^2 \sin^2 [\pi(\cos\psi - \cos\delta)\, d/\lambda] \tag{6.31}$$

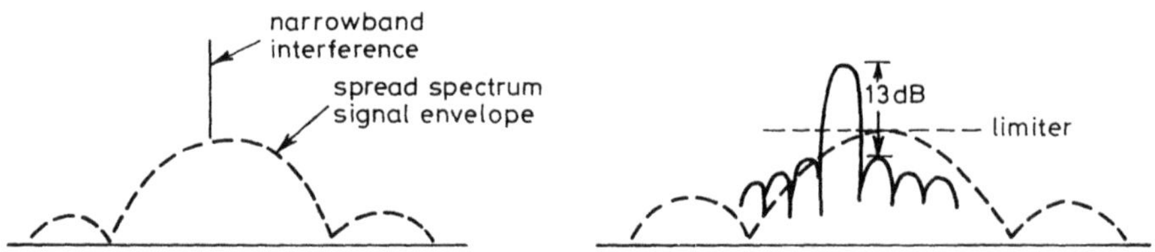

Fig. 6.12 *Narrowband interference before and after processing by transform circuitry*

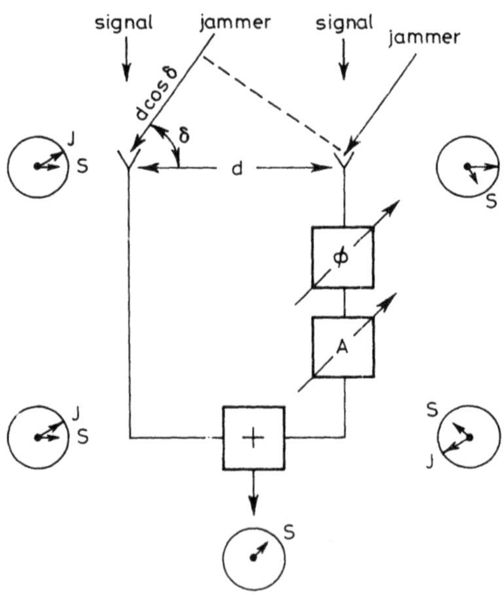

Fig. 6.13 *Null steering, basic principle*

Figure 6.15 shows the antenna diagram for different values of d and for $\delta = 90°$, $60°$ and $0°$. The diagram for $\delta = 0°$ gives a 'wide null' and indicate certain requirements with respect to antenna configuration adjustments.

Besides the antenna diagram itself, another point of great interest is the sensitivity of the null with respect to interference frequency. The antenna response as a function of frequency is given by

$$E^2 = 4E_1^2 \sin^2\left(\frac{\pi d}{\lambda} \cos \delta \frac{\Delta f}{f}\right) \quad (6.32)$$

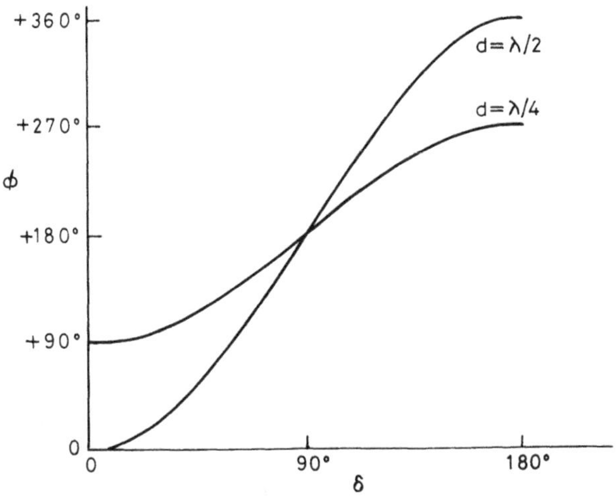

Fig. 6.14 *Connection between phase and direction to null position*

where f is the frequency to which the null is adjusted and Δf is the departure from this. Note that when $\delta = 90°$ the frequency variation does not influence the response.

An automatic processor controlled null steering system based on these principles is described in Section 6.5.3.

6.5.1. *Feedforward controlled adaptive band reject filter*

The concept for narrowband interference rejection treated here is based on a feedforward controlled adaptive band reject filter [3]. The idea behind the approach is to feed the received signal to a phase locked loop which uses the signal as a reference in estimating the interference frequency. Then a feedforward adjustment of a band reject filter is carried out. In order for the concept to work satisfactorily it is necessary to make demands on the precision of the estimated frequency. Assuming that the programming of the band reject filter can be made exact, a maximal tolerable error in estimated frequency Δf is given by

$$\Delta f < 0.5(BW_s BW_j)$$

where BW_s and BW_j are the widths of the stopband and interference respectively. In practice the frequency estimates are satisfactory as long as the phase loop is locked.

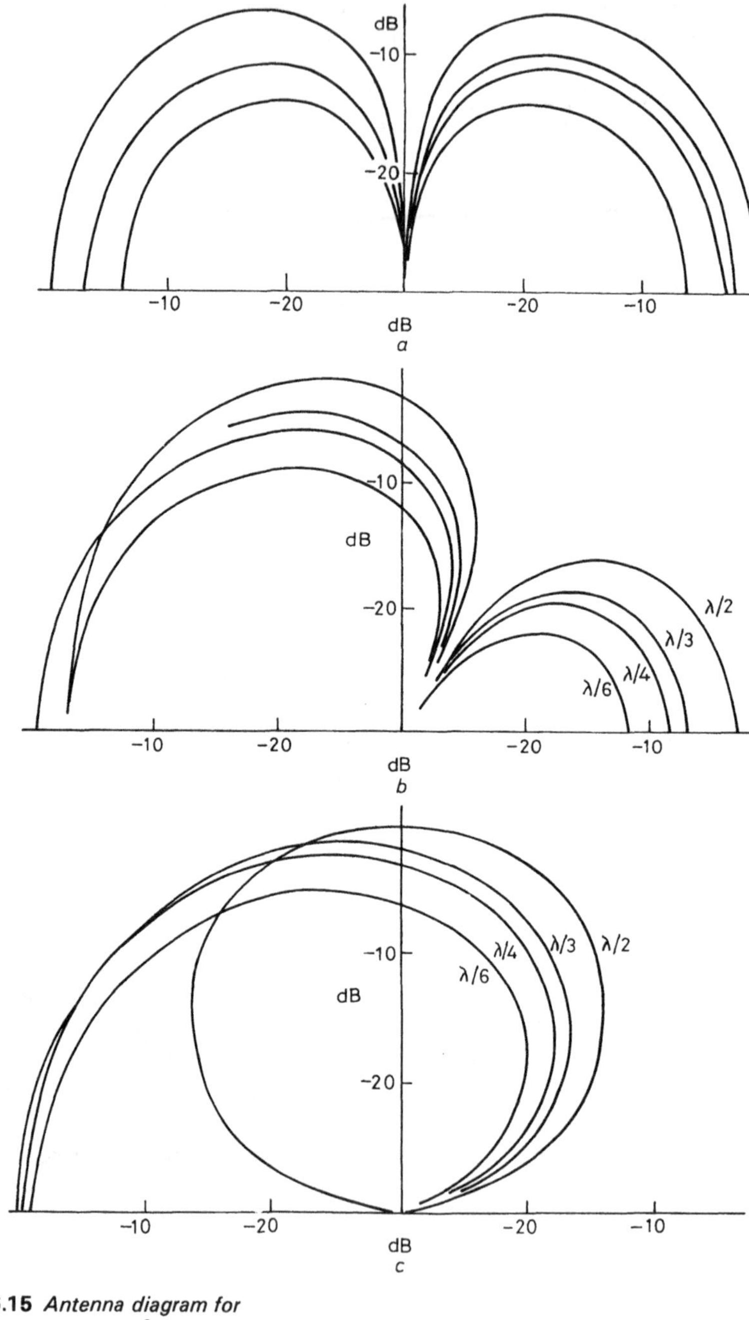

Fig. 6.15 *Antenna diagram for*

a $\delta_0 = 90^0$

b $\delta_0 = 60^0$

c $\delta_0 = 0^0$

The complete circuit then has three main elements – the adaptive band reject filter, a limiter and the phase locked loop frequency estimator. In reality it will be the quality of the band reject filter which will determine the degree of rejection. A compromise must be made in the conflict between large rejection and small bandwidths. The bandwidths should be kept at a minimum in order not to degrade the correlation peaks too much. Interference from conventional radiostations is likely to have a bandwidth of < 25 kHz.

The adaption of the band reject filter can be realized by heterodyning the received signal with a local oscillator to a fixed frequency matching the stopband filter response. Then after filtering the resulting signal can be mixed with the same local oscillator. The principle is shown in Fig. 6.16. The filter delay will put limits on how well the circuit can follow and reject an interferer sweeping across the spread spectrum band. If the circuit locks to such an interferer the local oscillator in the filter circuit may have changed during the signal path delay from one mixer to the other. This introduces frequency modulation.

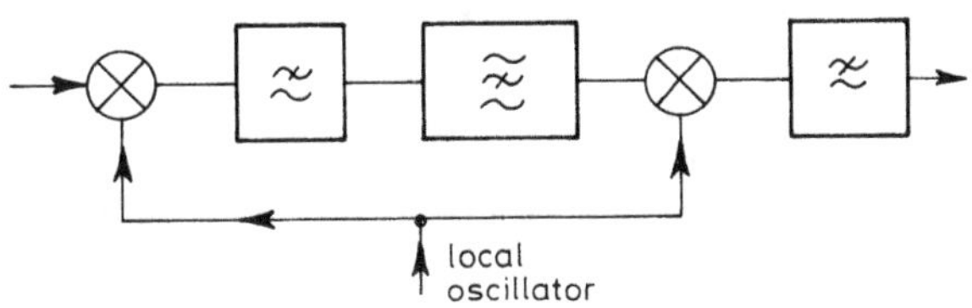

Fig. 6.16 *Heterodyning of received signal to match a band reject filter of fixed frequency*

The filter itself may be difficult to realize for small bandwidths and large rejection with conventional components. Filter may however be implemented with crystal resonator technology. A hard limiter will usually be required between the received signal and the phase locked loop, since most automatic gain control (AGC) circuits will not limit the signal dynamics to a level acceptable to the phase locked loop. The requirement is a constant envelope independent of the input level, which is not possible with linear AGC circuits. A possible solution is to make use of differential amplifiers with a gain large enough to limit the signal.

The central part of the rejection unit is the phase locked loop itself, estimating the interfering frequency. To match the circuit realizations discussed so far, the output from the phase locked loop must be a carrier centred at a frequency which will place the interferer inside the band reject width after heterodyning. If several narrowband interferers are present within the spread spectrum bandwidth it would usually be best if the circuit reliably locks to the strongest signal. The ability to lock to the strongest signal will require a minimum difference in interference signal levels, the magnitude of the minimum difference being dependent on the interference level compared with the kT_0B noise level and the spread spectrum signal. An additional constraint on the loop is that the acquisition and lockin bound must match the wideband spread spectrum receiver. To achieve an acquisition band of several megahertz might make it necessary to search for the narrowband interferers by sweeping the free running frequency of the voltage controlled oscillator through the band. The lockin time should be as short as possible if the interferers are expected to change frequency frequently.

A realization of an interference rejection unit based on the principles discussed has been built and tested. The circuit is shown in Fig. 6.17. The unit is designed for a VHF spread spectrum system with a 5 MHz bandwidth. The circuit achieves a mean lockin time of 1.5 ms for narrowband interferents inside the 5 MHz band and is able to operate in a multiple interferents situation. The rejection is better than 50 dB with a 10 kHz wide band reject filter. The band reject filter is implemented in crystal technology. This performance makes the circuit even able to follow certain frequency hopping radios. Fig. 6.18 shows the rejection capabilities for the circuit.

6.5.2 *CW interference suppression system using Fourier transform techniques*

The system is implemented using linear FM sweep signals and chirp filters [4] and is shown in Fig. 6.19. Interference reduction is achieved with limiting in the frequency domain. The system is designed for continuous spread spectrum signals with a bandwidth of 5 MHz.

The SAW filter used is a downchirp filter with a characteristic as shown in Fig. 6.20. The centre frequency is 60 MHz and the bandwidth is 20 MHz for a time of 10 μs.

Since the input signals might be time continuous, a data cycle of 100% is required for the transformation system. From Fig. 6.19 the first chirp signal must have a period equal to the time for analysis, giving the required frequency resolution

$$T_{\text{chirp}} = T_{\text{input}}$$

and the bandwidth becomes

$$B_{\text{chirp}} = 2|\mu| T_{\text{chirp}} = 2|\mu| T_{\text{input}}$$

Now the bandwidth of the SAW filter must match the bandwidth of the signal entering the filter. This becomes

$$B_{\text{filter}} = B_{\text{input}} + B_{\text{chirp}} = B_{\text{input}} + 2|\mu| T_{\text{input}}$$

The required length of the filter impulse response then becomes

$$\begin{aligned} T_{\text{filter}} &= \frac{B_{\text{filter}}}{2|\mu|} = \frac{B_{\text{input}}}{2|\mu|} + T_{\text{input}} \\ &= 2T_{\text{input}} = 2T_{\text{chirp}} \end{aligned}$$

Similarly,

$$B_{\text{filter}} = 2B_{\text{input}}$$

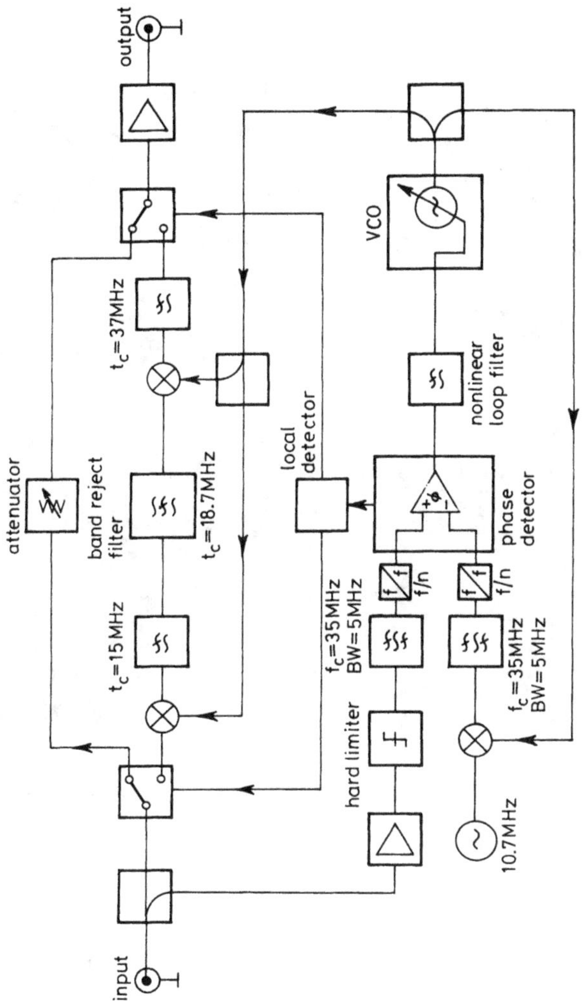

Fig. 6.17 *Block diagram showing feedforward adaptive band reject filter*

If the chirp signal is chosen with a bandwidth of 15 MHz, then $T_{chirp} = 7.5\,\mu s$ for a chirp rate of 2 MHz/μs matching the characteristic of the chirp filter. This gives a time for analysis of the input signal equal to 7.5 μs. The bandwidth of the signal entering the chirp filter now becomes 5 MHz + 15 MHz = 20 MHz, fully utilizing the 20 MHz bandwidth of the chirp filter.

Since the input signal is mixed with an upchirp signal before it enters the chirp filter, lower frequency components will arrive earlier than higher frequency components.

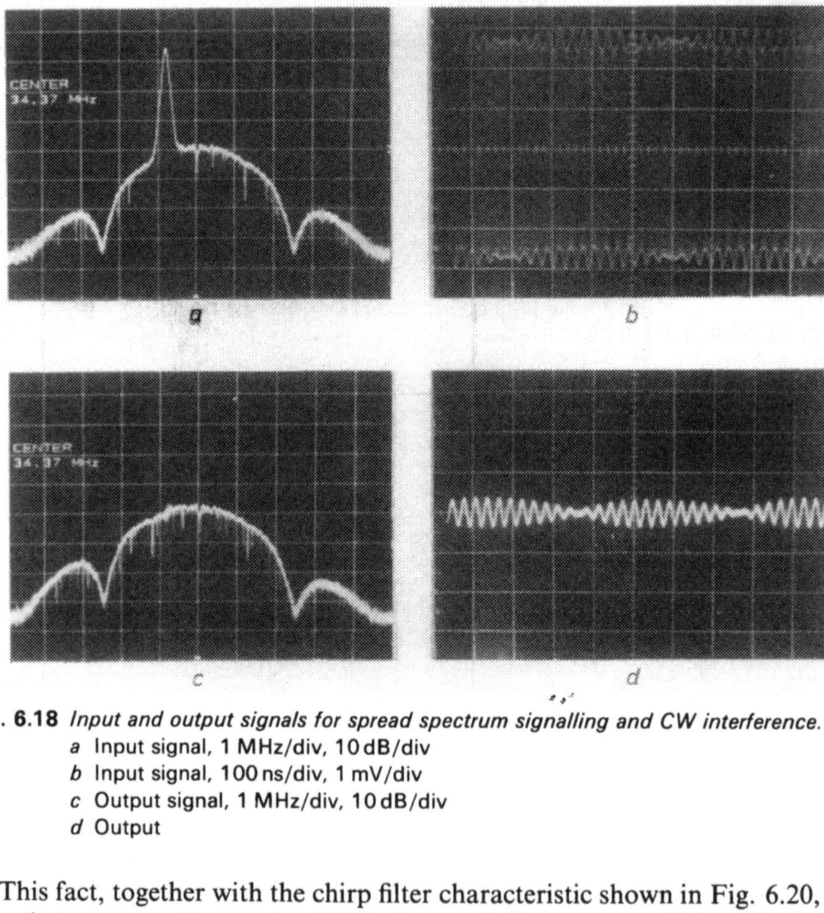

Fig. 6.18 *Input and output signals for spread spectrum signalling and CW interference.*
a Input signal, 1 MHz/div, 10 dB/div
b Input signal, 100 ns/div, 1 mV/div
c Output signal, 1 MHz/div, 10 dB/div
d Output

This fact, together with the chirp filter characteristic shown in Fig. 6.20, will give time compression, and a 100% duty cycle becomes possible with a 7.5 μs time for analysis. Then finally the length of the second chirp signal necessary for the transform is determined from the signal leaving the chirp filter. This gives a bandwidth equal to the input signal and a time equal to the time for analysis of the input signal.

Interference suppression with fixed thresholds is then accomplished using limiting amplifiers. Critical factors for the limiting network are the phase

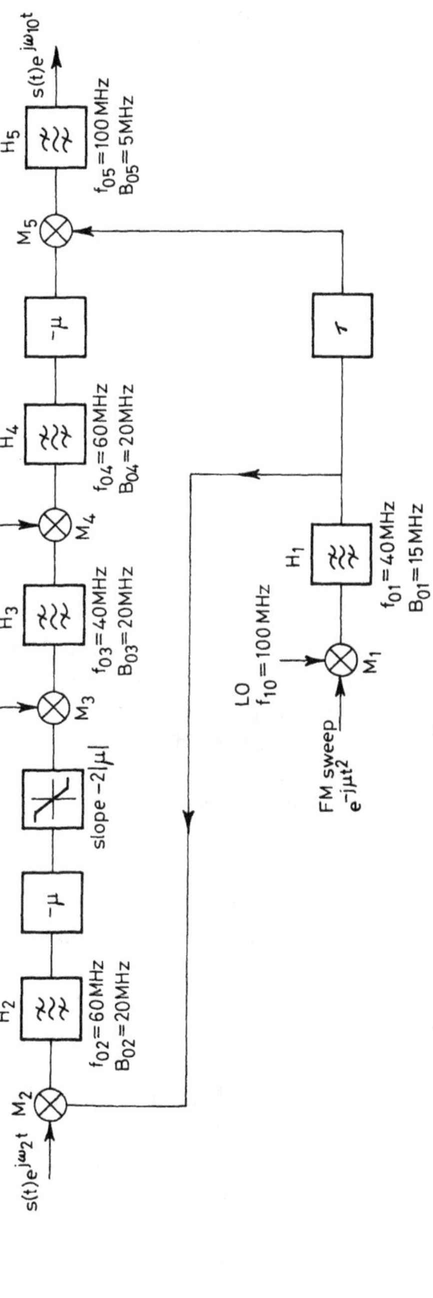

Fig. 6.19 *Circuitry for narrowband interference rejection using chirp-z transformation*

response and required linearity. The signal is then finally inverse Fourier transformed using two chirp signals and a SAW down chirp filter. Another critical factor appears at the last multiplication of the inverse transform. At this point the two signals multiplied must be synchronous, requiring the signal delay through the transformation path to be an integer number of 7.5 μs.

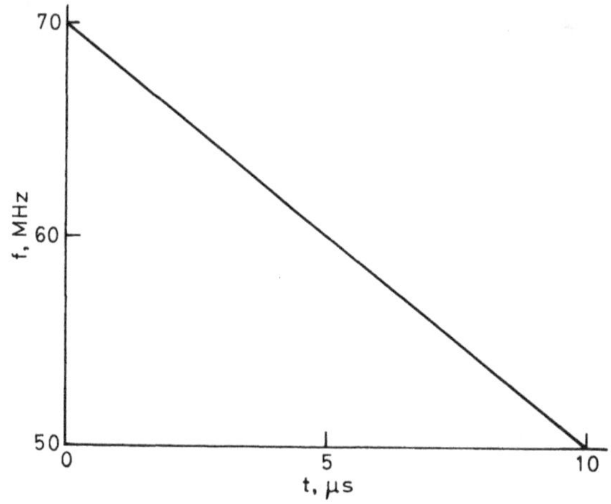

Fig. 6.20 *Downchirp filter characteristic*

Figs 6.21–6.24 shows the system in operation. Fig. 6.21 shows the bandpass filtered data signal. Fig. 6.22 shows the frequency spectrum of this spread spectrum signal and the narrowband interference with an *S*/*I* ratio of − 18 dB. In Fig. 6.23 the resulting data output is shown with no interference suppression. The *S*/*I* ratio is now about 0 dB. Then finally Fig. 6.24 shows the data output with the transformation and limiting in operation.

6.5.3 *Realization of steerable null antenna system*

The antenna network is based on the fundamental principle of Fig. 6.13, but the realization [5] makes it possible to adjust phase and amplitude independently as shown in Fig. 6.25. Only the amplitude is adjusted in the left-hand branch and only the phase in the right-hand branch. The switches S_1 and S_2 make it possible to change the polarity of X and Y in order to make all phases from 0° to 360° possible by just changing X and Y.

The required accuracy of the phase adjustments and the amplitude adjustments is related to the degradation the inaccuracies introduce. Table 6.3 shows the degradation for errors in amplitude and phase. The implemented system has a largest phase error of < 2° and a largest amplitude error < 3%. This should give little system impairment.

The null steering takes place at RF frequencies and this puts fairly strict requirements on the circuit components. The 90° hybrid circuit is particularly critical since an imbalance at this point will result in errors in the actual phase shift, the step size of the phase adjustments will increase and errors in the amplitude adjustments will be introduced. These factors may all degrade the system. In the system implemented an imbalance of maximum 2 dB is acceptable for the hybrid.

Table 6.3 *Limitations in suppression due to errors in phase and amplitude*

Amplitude errors only									
Error (dB)	0.08	0.17	0.225	0.34	0.42	0.8	1.58	2.5	3.0
Error (%)	1	2	3	4	5	10	20	33	41
Suppression (dB)	40	34	30	28	26	20	14	9.6	7.7

Phase errors only								
Errors (°)	:	1	2	3	4	5	10	20
Suppression (dB)	:	35	30	25	23	21	15	9

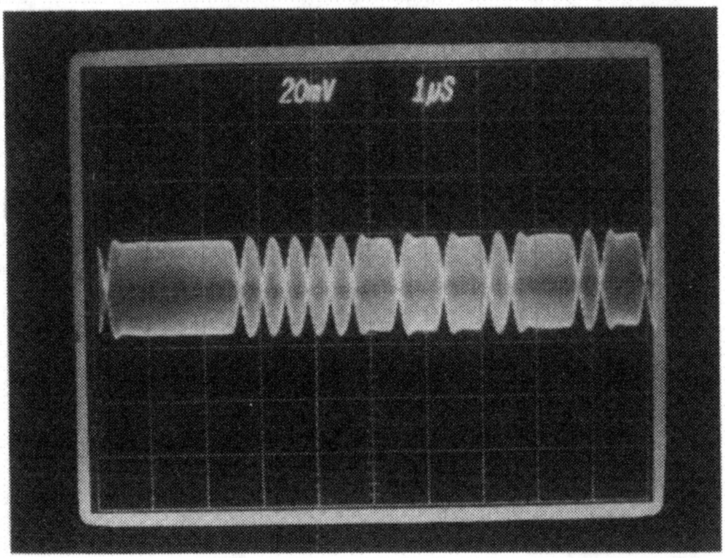

Fig. 6.21 *Bandpass filtered data signals*

An operator and a microprocessor are part of the overall antenna system. The basic philosophy of the operator is that he would make the decision that significant interference is being experienced and then initiate the antenna nulling system. To be able to insert a null in the right direction, the operator/processor requires information of the interference power after the summation network. This is accomplished using envelope detection at an intermediate frequency and an A/D converter connecting the lowpass filtered signal and the microprocessor (or display). The processor must be able to remember the phase/amplitude position giving the smallest signal energy and then go back to that position after a complete 360° search.

The system is only able to nullify one interference source, and from the

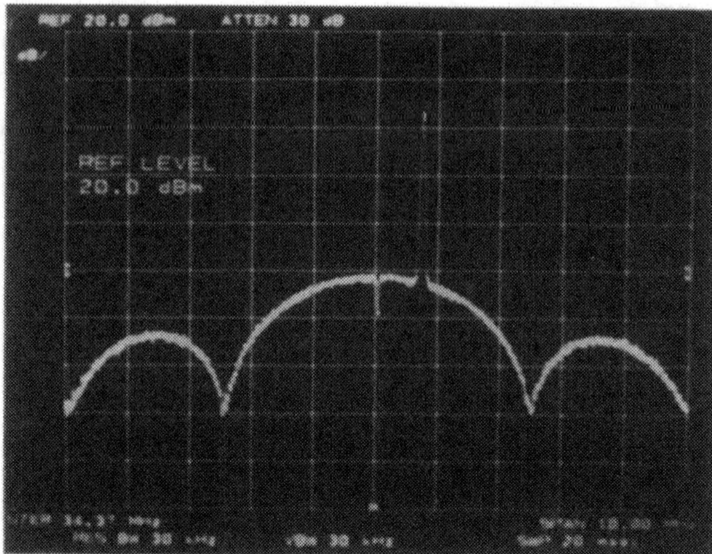

Fig. 6.22 *Spread spectrum signal with narrowband interference, S/I = −18 dB*

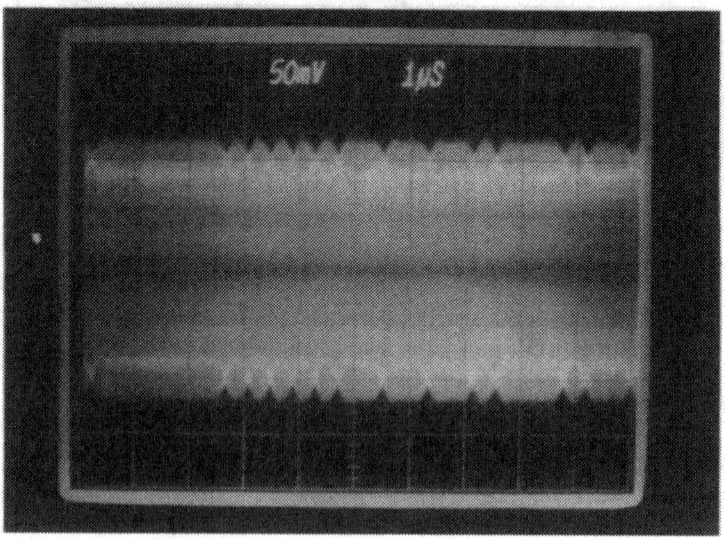

Fig. 6.23 *Data output with S/I = −18 dB and no interference suppression*

previous discussion it is obvious that the null will be directed towards the strongest narrowband interference. Such a simple algorithm for interference suppression is possible because spread spectrum modulation is used and the

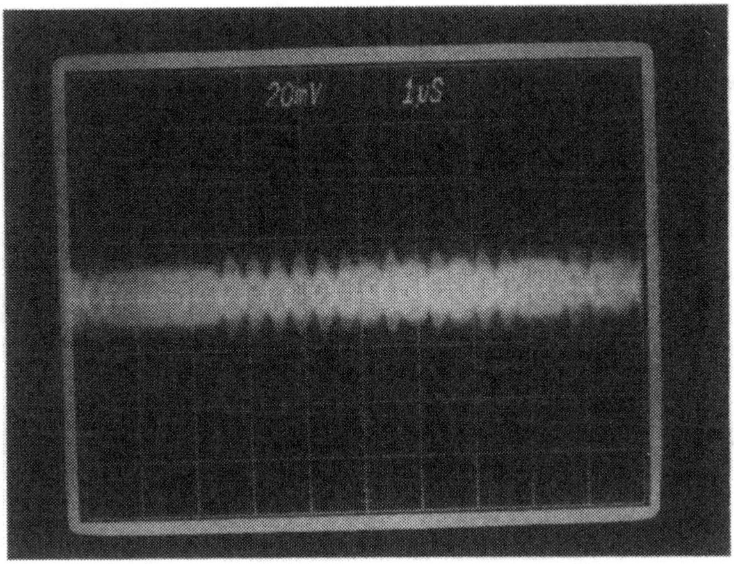

Fig. 6.24 *Data output with S/I = −18 dB and limiting in the frequency domain*

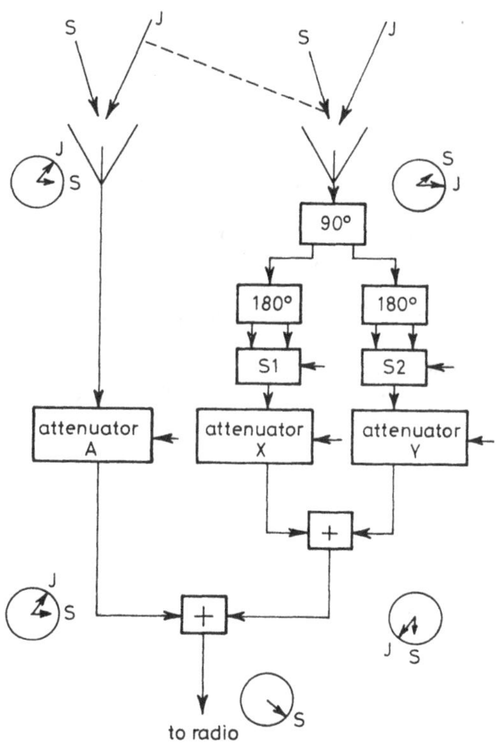

Fig. 6.25 *Realization of amplitude and phase adjustments in antenna null steering system*

wanted signal is usually received with negative signal/noise ratios. However, cases do appear for which suppression of the wanted signal may occur, but this is in practice solved by certain operational and configurational procedures. Fig.

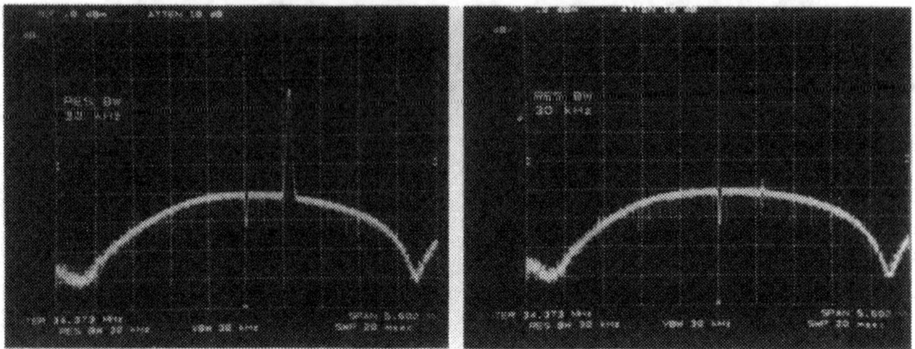

a b

Fig. 6.26 *Null steering antenna network.*
a Unsuppressed narrowband interference
b Antenna null directed toward interference

6.26 shows the system in operation. First the unsuppressed narrowband interference is seen inside the spread spectrum bandwidth. Then the null is directed toward the interference, and this is seen to be greatly reduced.

References

1 ROBINSON, J. L., and SKAUG, R.: 'A practical approach for the operation of HF portable data networks'. *IEEE Conf. Publ. 206*, 1982, pp. 81–85

2 SUSSMAN, S. M., and FERRARI, E. J.: 'The effects of notch filters on the correlation properties of a PN signal', *IEEE Trans.*, 1974, **AES-10**, pp. 385–390.

3 NORLAND, M. O.: 'Feed-forward controlled adaptive bandstop filter for interference rejection in spread spectrum communication systems' (in Norwegian), Diploma Thesis, Norwegian University of Technology, Trondheim, May 1983.

4 KLIPPENBERG, N.: 'Interference rejection in a direct sequence spread spectrum system' (in Norwegian), Diploma Thesis, Norwegian University of Technology, Trondheim, May 1983.

5 THORVALDSEN, T.: Internal Memorandum, NDRE.

Chapter 7

Final word

The theory of spread spectrum communication has been known for a very long time, and work on direct sequence spread spectrum (DSSS) systems can be traced back to the 1950s. Nevertheless, it has taken more than 20 years for the technique to emerge from the laboratory and gain more than a very limited applicability. Both for military applications in field equipment, and for commercial applications such as mobile telephone and microwave communication, size, power consumption and cost have been the main reasons why spread spectrum modulation has not gained ground faster. Another important obstacle, particularly for the civil systems, has been the lack of any international agreements on frequency provisions for wideband systems.

As discussed in Chapter 4, technological development in recent years has been impressive. The matched filter has been and still is the major cost component in a DSSS system; its size and power consumption have, however, been dramatically reduced. Fig. 7.1 shows a 64 bit correlator from the early 1970s made up of standard TTL circuitry, and a correlator from the early 1980s custom designed with silicon on sapphire MOS technology. The former makes up three 'Double Europe' printed circuit boards and uses 15 W of power; the latter measures 10 mm by 10 mm and uses 15 mW of power.

The price has unfortunately not been reduced in the same manner. In order to reduce cost significantly it is still essential to go thoroughly through the requirements for processing gain and number of quantization levels before deciding on length and number of correlators. The nondigital matched filter, like the SAW convolver described in Chapter 4, offers a very compact solution, but because of the large internal attenuation requires a carefully designed amplification network. The high speed of the signal propagating inside the convolver also requires well designed high speed circuitry to supply necessary codes and timing for synchronization.

No doubt the emergence of technology capable of operating at constantly higher speeds and at constantly lower prices will enable more elegant processing algorithms. This will reduce the hardware size and cost so that the matched filter will no longer be a major cost factor in a radio set. This will in turn lower the threshold at which spread spectrum modulation becomes appropriate.

In this book we have limited the discussion to spread spectrum systems with a burst signalling format, and in particular we have addressed the importance of spread spectrum to military communication systems. The majority of spread spectrum systems are still military, the main reason for this being, as mentioned, the growth in cost for systems with this modulation. For the military sector the

Fig. 7.1 *Development of technology for correlator design. Shows a 1970 version of a 64 bit correlator designed at NDRE from standard TTL circuitry and a 1980 version of the same correlator designed at NDRE in SOSMOS technology*

urgent ECM threat has rendered necessary immediate improvements, even costly ones. However, it should be quite obvious that the multiple access possibilities and the multipath and interference rejection capabilities described in Chapters 3, 5 and 6 could be just as advantageous to civilian systems.

In recent years there has been particular interest in spread spectrum modulation, from the viewpoint of spectral efficiency, for mobile communication for both military and civilian use. It was shown in Chapter 6 that for mobile communication DSSS techniques may run into special problems relating to the large geographic areas to be covered and the use of signals occupying the same frequency band – the near–far problem. However, the use of a wideband system results in a form of frequency diversity that significantly reduces the degradation in performance that normally arises from rapid fading. If effective power control in order to prevent users near the base station from overpowering more distant users is not achievable, then multiple access methods based on orthogonal convolutional coding with on/off signalling and Viterbi decoding may be a solution. These schemes can solve the near–far problem and enable the users to access the system at any time without waiting for a free channel. There will be no hard limit to the number of active users, and if the number exceeds the design value the result is a graceful degradation and not denial of access. Since each potential user of the system can be assigned a unique spreading code, message privacy can be achieved.

Another area where both civilian and military use are foreseen is HF skywave communication. The multipath rejection capability of DSSS systems makes the modulation of special interest for this frequency band. In Canada [1] an HF modem using spread spectrum techniques has been developed to establish links to and from remote areas under conditions of multipath fading and narrowband interference. Bit error rate performances were reported to be improved compared with conventional FSK modems by orders of magnitude under conditions of heavy selective fading.

Although technological development has contributed to the increase in the future potential and applicability of spread spectrum systems, the lack of agreements on frequency provision for these wideband systems is still a major obstacle. And the problem is one not only of agreements but also of knowledge of the interaction between conventional systems and DSSS systems. The DSSS system will have to share frequency assets with conventional systems, creating problems both for the existing systems and the future DSSS system. Action aimed at providing for accommodation of future wideband systems is essential. To establish future sharing patterns, much more information is needed on the interaction and interference between the systems for all the different frequency bands. To many, the introduction of wideband systems must be seen as a paradox at a time when frequency assets are scarce and every possible effort is made to husband them, for example by reducing the spectral occupancy of digital speech. However, the introduction of DSSS systems must be integrated with techniques to reduce baseband occupancy and a serious reconsideration of what information is really essential. In DSSS systems information rate has been traded against security for jamming, interference and low probability of intercept. Generally any reduction in baseband occupancy will increase the possible system processing gain.

References

1 CHOW, I., CAVERS, K., and LEE, P. E.: 'A spread spectrum modem for reliable data transmission in the high frequency band'. *IEEE Conf. Publ. 206*, 1982, pp. 125–127

Appendix A

Expression for crosscorrelation between *m*-sequences using their autocorrelation functions

In order to relate the crosscorrelation of sequences to their autocorrelation, one might apply the following theorem due to Gold [1]:

$$\sum_{l=0}^{p-1} [\theta^{kr}(l)]^2 = \sum_{l=0}^{p-1} \left[\sum_{i=0}^{p-1} a^k(i + l)a^r(i) \right]^2$$

from the definition of the crosscorrelation $\theta^{kr}(l)$. The right-hand side can then be developed to give

$$= \sum_{l=0}^{p-1} \sum_{j=0}^{p-1} \sum_{i=0}^{p-1} a^k(i + l)a^r(i)a^k(j + l)a^r(j)$$

$$= \sum_{j=0}^{p-1} \sum_{i=0}^{p-1} a^r(i)a^r(j) \sum_{l=0}^{p-1} a^k(i + l)a^k(j + l)$$

$$= \sum_{j=0}^{p-1} \sum_{i=0}^{p-1} a^r(i)a^r(j)\theta^k(j - i)$$

$$= \sum_{i=0}^{p-1} \sum_{l=0}^{p-1} a^r(i)a^r(l + i)\theta^k(l)$$

$$= \sum_{l=0}^{p-1} [\theta^r(l)][\theta^k(l)] \tag{A.1}$$

From this it is possible to develop the result [2]

$$\sum_{l=1-p}^{p-1} [C^{kr}(l)]^2 = \sum_{l=1-p}^{p-1} [C^k(l)][C^r(l)]$$

$$= p^2 + 2 \sum_{l=1}^{p-1} [C^k(l)][C^r(l)] \tag{A.2}$$

for sequences where $C^k(0) = C^r(0) = p$.

Then

$$\sum_{l=0} [C^{kr}(l - p)]^2 + [C^{kr}(l - p + 1)]^2 + [C^{kr}(l)]^2 + [C^{kr}(l + 1)]^2$$

$$= 2 \sum_{l=1-p}^{p-1} [C^{kr}(l)]^2$$

$$= 2_p^2 + 4 \sum_{l=0}^{p-1} [C^k(l)][C^r(l)] \qquad \text{(A.3)}$$

References

1 GOLD, R.: 'Maximal recursive sequences with 3-valued recursive cross-correlation functions', *IEEE Trans.*, 1968, **IT-14** (1), pp. 154–156

2 PURSLEY, M. B., and SARWATE, D. V.: 'Evaluation of correlation parameters for periodic sequences', *IEEE Trans.*, 1977, **IT-23** (4), pp. 508–513

Appendix B

Crosscorrelation parameters for maximal length sequences

Let $\boldsymbol{a}^k$ be a maximal length sequence with period $p = 2^m - 1$ and $\boldsymbol{a}^r$ another m-sequence with the same period.

Both sequences thus have an autocorrelation given by

$$\begin{aligned}\theta^k(l) &= 1 & l &= 0 \bmod m\\ \theta^r(l) &= 1 & l &= 0 \bmod m\\ \theta^k(l) &= -1/p & l &\neq 0 \bmod m\\ \theta^r(l) &= -1/p & l &\neq 0 \bmod m\end{aligned}$$

The crosscorrelation between $\boldsymbol{a}^k$ and $\boldsymbol{a}^r$ is given by

$$\theta^{kr}(l) = \frac{1}{p}\sum_{s=1}^{p} a_s^k a_{s+l}^r$$

The mean $\overline{\theta^{kr}}$ of the crosscorrelation for all shifts is given by

$$\overline{\theta^{kr}} = \frac{1}{p}\sum_{l=1}^{p}\theta^{kr}(l) = \frac{1}{p^2}\sum_{l=1}^{p}\sum_{s=1}^{p} a_s^k a_{s+l}^r = \frac{(-1)(-1)}{p^2} = \frac{1}{p^2}$$

The variance σ^2 of the crosscorrelation for all shifts is given by

$$\begin{aligned}\sigma^2 &= \frac{1}{p}\sum_{l=1}^{p}[\theta^{kr}(l)]^2 - (\overline{\theta^{kr}})^2\\ &= \frac{1}{p^3}\sum_{l=1}^{p}\sum_{s=1}^{p}\sum_{t=1}^{p} a_s^k a_t^k a_{s+l}^r a_{t+l}^r - \frac{(-1)^2(-1)^2}{p^2p^2}\end{aligned}$$

Summing with respect to l:

$$= \frac{1}{p^2}\sum_{s=1}^{p}\sum_{t=1}^{p} a_s^k a_t^k \theta^r(s-t) - \frac{1}{p^4}$$

$$= \frac{1}{p^2}\sum_{s=1}^{p}\sum_{t=1}^{p} a_s^k a_t^k\left[\frac{p-(-1)}{p}\delta_{s-t} + \frac{(-1)}{p}\right] - \frac{1}{p^4}$$

This is because $\theta^r(l)$ has period p and one peak during the period

$$\sigma^2 = \frac{1}{p^2}\frac{p-(-1)}{p^2}\sum_{s=1}^{p} a_s^k a_s^k + \frac{1}{p^2}\frac{(-1)}{p}\sum_{s=1}^{p}\sum_{t=1}^{p} a_s^k a_t^r - \frac{1}{p^4}$$

Summing with respect to s:

$$\sigma^2 = \frac{1}{p}\frac{p-(-1)}{p} + \frac{1}{p^2}\frac{(-1)}{p} - \frac{1}{p^4}$$

$$= \frac{p+1}{p^2} - \frac{1}{p^3} - \frac{1}{p^4}$$

For $p \gg 1$,

$$\sigma_2 = \frac{1}{p}$$

Appendix C

Bounds on odd autocorrelation and crosscorrelation of sequences

From the theorem of Gold [1] developed in Appendix A, it is also possible to show [2] that the odd correlation parameters are related through

$$\sum_{l=0}^{p} [\hat{\theta}^{kr}(l)]^2 = \sum_{l=0}^{p-1} [\hat{\theta}^{k}(l)][\hat{\theta}^{r}(l)] \tag{C.1}$$

Applying this to all members of a set A consisting of K sequences of period p where the inphase autocorrelation $\theta^k(0) = p$ for all $k \in A$, one obtains

$$\sum_{k\in A}\sum_{r\in A}\sum_{l=0}^{p-1} [\hat{\theta}^{kr}(l)]^2 = \sum_{\substack{k\in A\\ k\neq r}}\sum_{r\in A}\sum_{l=0}^{p-1} [\hat{\theta}^{kr}(l)]^2 + \sum_{k\in A}\sum_{l=0}^{p-1} [\hat{\theta}^{k}(l)]^2$$

Thus from (C.1):

$$\begin{aligned}\sum_{\substack{k\in A\\ k\neq r}}\sum_{r\in A}\sum_{l=0}^{p-1} [\hat{\theta}^{kr}(l)]^2 + \sum_{k\in A}\sum_{l=0}^{p-1} [\hat{\theta}^{k}(l)]^2 &= \sum_{k\in A}\sum_{r\in A}\sum_{l=0}^{p-1} [\hat{\theta}^{k}(l)][\hat{\theta}^{r}(l)]\\ &= \sum_{l=0}^{p-1}\left[\sum_{k\in A}\hat{\theta}^{k}(l)\right]^2\\ &= K^2p^2 + \sum_{l=1}^{p-1}\left[\sum_{k\in A}\hat{\theta}^{k}(l)\right]^2\end{aligned}$$

The left side of this equation is upper bounded by

$$K(K-1)p[\hat{\theta}^{kr}_{\max}(l)]^2 + Kp^2 + K(p-1)[\hat{\theta}^{k}_{\max}(l)]^2 \qquad 0 \leqslant l \leqslant p-1$$

The right-hand side is lower bounded by K^2p^2:

$$K(K-1)p[\hat{\theta}^{kr}_{\max}(l)]^2 + Kp^2 + K(p-1)[\hat{\theta}^{k}(l)]^2 > K^2p^2$$

$$[\hat{\theta}^{kr}_{\max}(l)]^2K(K-1)p + [\hat{\theta}^{k}_{\max}(l)]^2K(p-1) > p^2K(K-1)$$

$$\frac{[\hat{\theta}^{kr}_{\max}(l)]^2}{p} + \frac{[\hat{\theta}^{k}_{\max}(l)]^2}{p}\,\frac{p-1}{p(K-1)} > 1$$

References

1 GOLD, R.: 'Maximal recursive sequences with 3-valued recursive cross-correlation functions', *IEEE Trans.*, 1968, **IT-14** (1), pp. 154–156

2 SARWATE, D. V.: 'Bounds on crosscorrelation and autocorrelation of sequences', *IEEE Trans.*, 1979, **IT-25** (6), pp. 720–725

Appendix D

Experiments with spread spectrum modulation over dispersive medium

To provide some experimental data on the useful bandwidth for spread spectrum systems, test equipment consisting of a spread spectrum 2-PSK transmitter was constructed [1]. The receiver consisted of a spread spectrum passive correlator for direct sequence modulation (Fig. D.1). The passive correlator consists of two lowpass inphase and quadrature filters implemented as two identical digital matched filters constructed to match the spreading code used. The receiver was designed to offer linear phase response to the variable signal bandwidths. The linear phase requirement emerged from the wish to measure the nonlinearities of the ionosphere undistorted by the test equipment.

Two analogue filter banks, allowing the bandwidth to be varied interface the radioset and the spread spectrum modem.

The bandwidth of the spread spectrum signal is proportional to the clock rate of the spreading code. The equipment constructed enables a choice of five bandwidths of approximately 2.5 kHz, 10 kHz, 40 kHz, 80 kHz and finally 160 kHz.

After processing every sequence period is presented as a clearly distinguishable signal peak. Between these peaks the signal level will in principle be almost zero. However, if the received signal contains multipath contributions these will show up as signal peaks in between the main peaks.

The magnitude of the signal peak will depend on the received signal strength and the degree to which the signal waveform is distorted through the propagation medium. The relative multipath delays can be read off directly on an oscilloscope display. The resolution of multipath delays is determined by the signal bandwidth, and the equipment described offers a resolution varying from 0.8 ms to 12.8 μs for the bandwidth range. The analogue/digital converters operating at the input of the matched filters have 5 bit resolution and, because no automatic gain control is in operation, the relative strength of the multipaths can be viewed on the display.

The advantage of using a matched filter is that the system then in principle becomes asynchronous. The important tasks for the receiver are to detect the correlation peaks and derive data sampling time information. A control unit

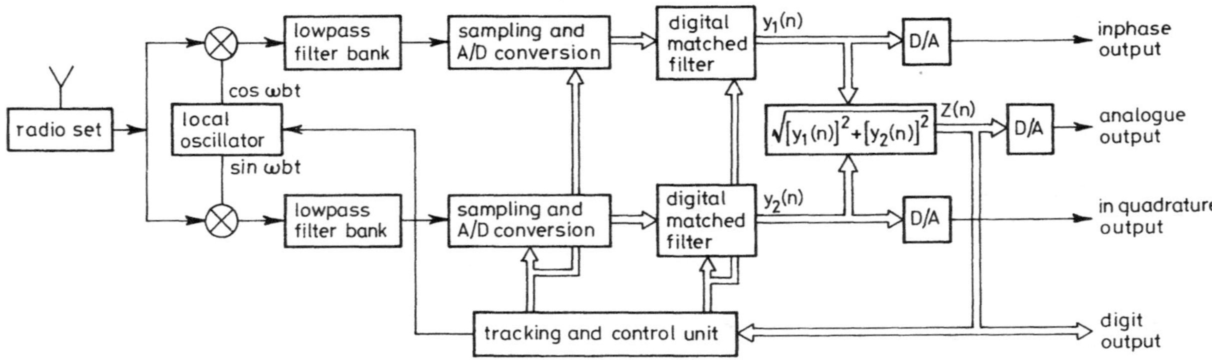

Fig. D.1 *Receiver correlator implemented as digital matched filters*
(SKAUG, R., 'Experiments with spread spectrum modulation on radiowaves reflected from the ionosphere', *Arch. Elektron and Uebertragungstech*, 1981, **35** (4), pp. 151–155.
Copyright © AEU)

connected to the output from the correlator is used to obtain maximum correlation, adjusting the time of sample at the A/D converters. For multipath the individual modes will tend to exhibit independent variations in amplitude and phase, and therefore at certain times the dominant propagation mode may change. This will have to be reflected in any receiver tracking circuitry.

The tracking circuitry used provides flexibility by allowing the user to set manually a number of variables important to the tracking function.

One of these variables was the time to elapse from the 'disappearance' of a multipath component until a search routine for a new multipath component was initiated. The user also controlled different threshold levels and a parameter determining the number of times a detection threshold should be exceeded before the receiver acknowledged a useful signal and locked to a particular multipath component.

D.1 Test procedure

Test transmissions took place over a 840 km path between Bodø (67°N, 14°E) and Kjeller (60°N, 11°E) on 8.1 MHz.

Measurement periods were chosen both at times when 8.1 MHz was sufficiently far below the predicted maximum usable frequency (MUF) to make multihop modes probable, and at times when 8.1 MHz was very close to the MUF.

A counter was connected to the spread spectrum receiver output and a threshold level was set to a value giving a small number of false alarms. Then the number of correlation peaks above this threshold level were counted and compared with the theoretical total number of possible peaks, resulting in a detection percentage.

D.2 Basic results

The measurements were carried out under a variety of propagation conditions, and some selected field test data can be presented.

D.2.1 *Carrier frequency below predicted maximum usable frequency*

The measurements were recorded between 12.00 and 13.00 UT when 8.1 MHz was sufficiently below the predicted maximum usable frequency to make it possible that both one- and two-hop modes would be observed.

Several recordings show the presence of low angle rays of two modes (1F and 2F). However, in addition two magneto-ionic components were occasionally observed.

The test frequency of 8.1 MHz was close to the MUF for a path between Bodø and the midpoint to Kjeller and thus the MUF for a 2F path. A typical situation is illustrated in Fig. D.2*a*. The weaker 2F path is delayed approximately 1.1 ms with the 2F ordinary and extraordinary components separated by approximately 150 μs to 200 μs. Fig. D.2*b* shows the performance of the five bandwidths.

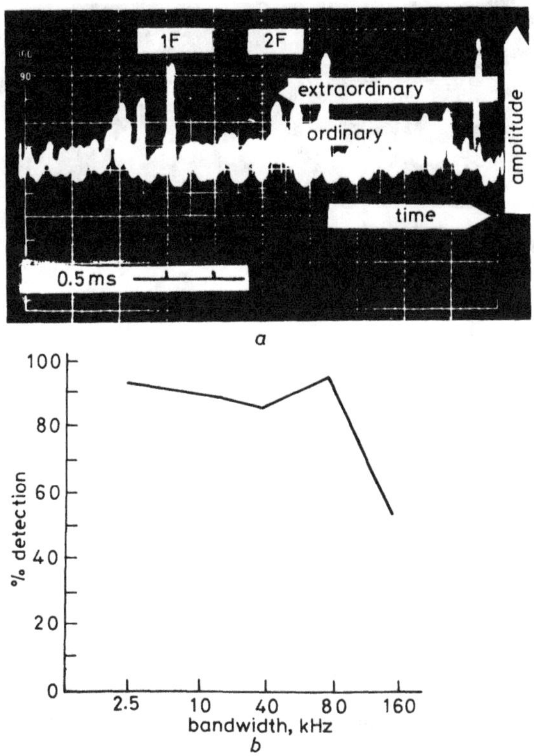

Fig. D.2 *a* Correlator output recording
b Comparison of the performance for five different bandwidths
(SKAUG, R., 'Experiments with spread spectrum modulation on radiowaves reflected from the ionosphere', *Arch. Elektron and Uebertragungstech*, 1981, **35** (4), pp. 151–155.
Copyright © AEU)

For the situation exhibited in Fig. D.2*a* all five bandwidths available to the spread spectrum receiver were able to resolve the 2F path, since the quantity $\tau W > 1$ for all W, where τ is the differential time delay between 1F and 2F modes and W is the spread spectrum bandwidth. The multipath would on this occasion not greatly affect the performance for any of the test bandwidths. However, as illustrated in Fig. D.2*b*, a significant drop in performance was observed for the larger bandwidth of 160 kHz.

The situation illustrated in Fig. D.3*a* was obtained from measurements 40 days later. The sample differs primarily in showing the presence of an echo delayed approximately 0.65 ms and whose identity is not certain. The delayed path might be an M mode. On the other hand the 2F multipath delayed by $\approx$1.1 ms present in the sample shown in Fig. D.2*a* is not observed on this particular recording. An unidentified component is sometimes present with a delay of 25 μs to 50 μs. In Fig. D.3*b* the performance for the five bandwidths is again shown.

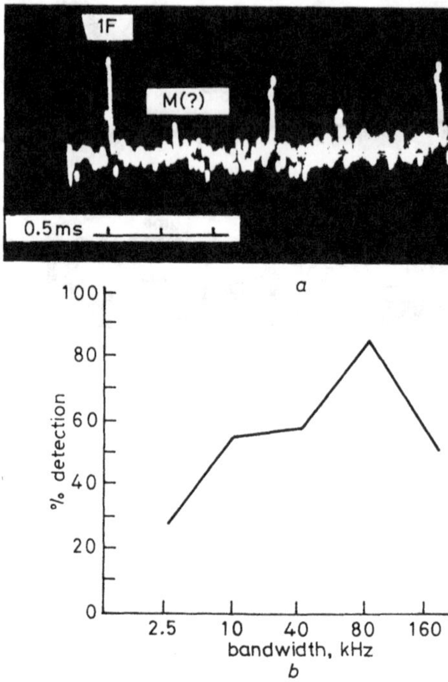

Fig. D.3 *a* Correlator output recording
b Comparison of the performance for five different bandwidths
(SKAUG, R., 'Experiments with spread spectrum modulation on radiowaves reflected from the ionosphere', *Arch. Elektron and Uebertragungstech*, 1981, **35** (4), pp. 151–155.
Copyright © AEU)

For this situation the quantity $\tau W < 1$ for the smaller 2.5 kHz spread spectrum bandwidth. The portion of the delayed aperture will produce a randomly phased component and this may contribute negatively to the correlation from the 1F path.

Fig. D.3*b* clearly illustrates the expected poor performance at the 2.5 kHz bandwidth. In fact the performance is continuously improved by increasing the bandwidth to 80 kHz. This could be explained by an unidentified mode which is resolved only for bandwidths 80 kHz. Again a drop in performance results when the bandwidth is increased to 160 kHz.

D.2.2 *Carrier frequency close to predicted maximum usable frequency*
Some measurements were recorded in late evenings or early mornings when the 8.1 MHz carrier frequency was close to the predicted maximum usable frequency.

As expected no multihop paths are observed in these data, but four paths were present, identified as 1F low and high angle modes, each with the two magneto-ionic components significantly separated.

A typical recording shows the low and high angle modes separated by approximately 0.8 ms and the magneto-ionic components by approximately 50 μs to 100 μs. The recording is shown in Fig. D.4*a*. The performance of the five bandwidths during this situation is shown in Fig. D.4*b*.

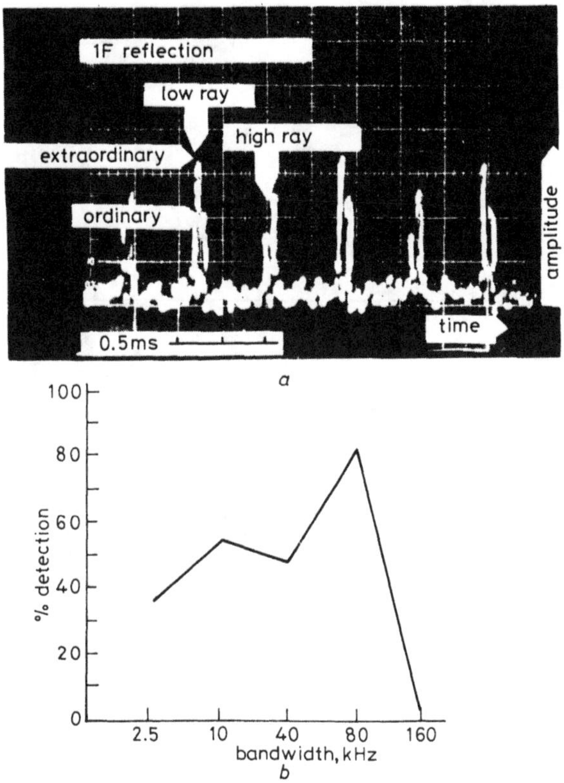

Fig. D.4 *a* Correlator output recording
b Comparison of the performance for five different bandwidths
(SKAUG, R., 'Experiments with spread spectrum modulation on radiowaves reflected from the ionosphere', *Arch. Elektron and Uebertragungstech*, 1981, **35** (4), pp. 151–155.
Copyright © AEU)

Once again the quantity $\tau W < 1$ for the smaller bandwidths, and the results are poor for the smaller bandwidths and for the 160 kHz bandwidth.

D.3 Discussion of results

Seven emdian values for midday measurements and four median values for morning measurements are shown in Fig. D.5. The plots clearly show the general poor performance for the largest 160 kHz bandwidth but comparable results for the remaining four bandwidths. However, measurements performed on 27 November are illustrated in Fig. D.3*b* and show the improvement obtained by increasing the bandwidth from 2.5 kHz to 80 kHz. Measurements done on 6 and 8 November differ from the rest in that there are no significant performance degradation for any of the bandwidths.

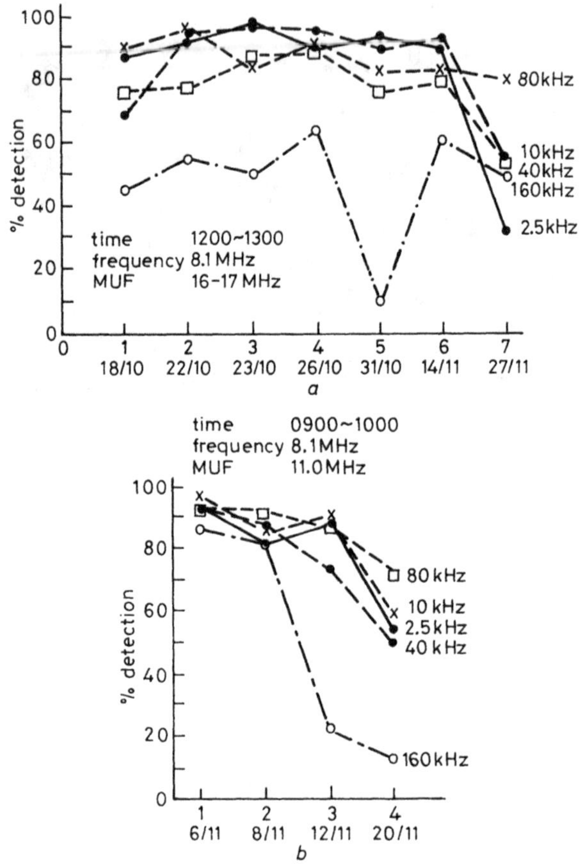

Fig. D.5 *a* Correlator output recording
b Comparison of the performance for five different bandwidths
(SKAUG, R., 'Experiments with spread spectrum modulation on radiowaves reflected from the ionosphere', *Arch. Elektron and Uebertragungstech*, 1981, **35** (4), pp. 151–155.
Copyright © AEU)

In addition to multipath interference the use of HF frequencies introduce a large amount of other user interference. This becomes increasingly serious as the bandwidth is increased. This is illustrated in Fig. D.6. In fact to make sure that the degradation usually encountered for the 160 kHz bandwidth was not due to an increased number of other users interfering with the test signal, the spread spectrum waveform was generated locally and added to the received 160 kHz noise spectrum. However, with this arrangement the performance for the

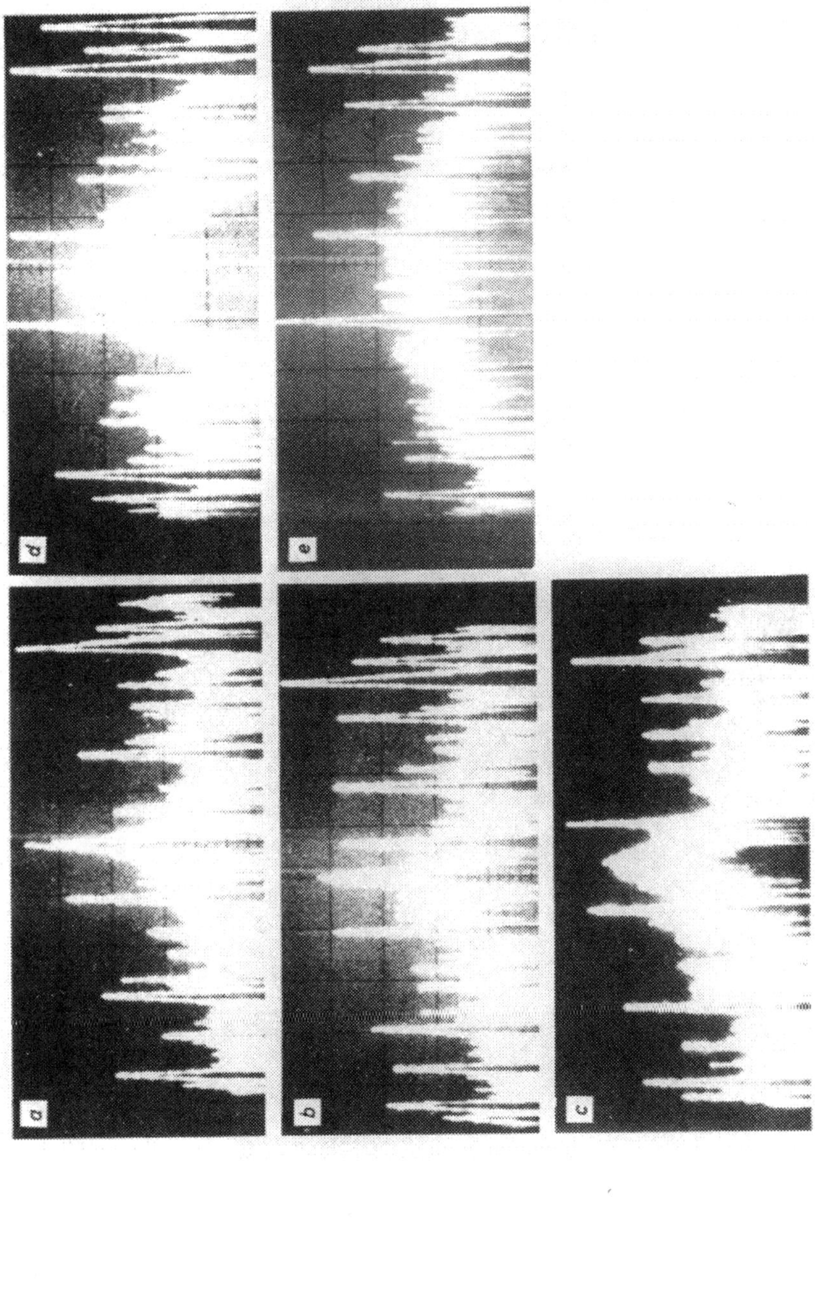

Fig. D.6 *RF spectrum at 8.1 MHz for five different spread spectrum bandwidths. a* 2.5 kHz; *b* 10 kHz; *c* 40 kHz; *d* 80 kHz; *e* 160 kHz

160 kHz bandwidth improved significantly, and compared well with the expected interference rejection capabilities of the system. A certain decorrelation between the interferers when it came to fading cycle was also observed. Thus the interferers did not necessarily reach their maximum level at the same time. To count the number of other users within the spectrum utilized by the spread spectrum system, and add them based on their maximum power level, is probably not a realistic way to describe the interference condition.

The persistent drop in performance for the 160 kHz bandwidth could not be traced back to increased interference, and thus requires some other explanation.

Some degree of scattering occurs whenever a radio wave is reflected from the ionosphere. The received signal may be regarded as possibly having a plane wave or specular component together with a diffracted component, which is composed of wavelets scattered with random phases from the ionosphere.

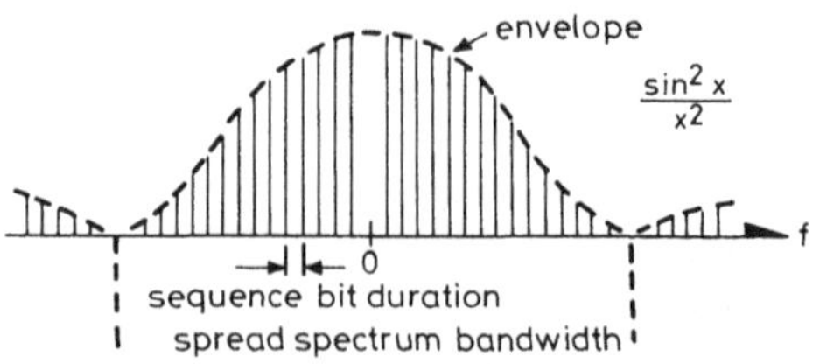

Fig. D.7 *Output spectrum for test transmitter*

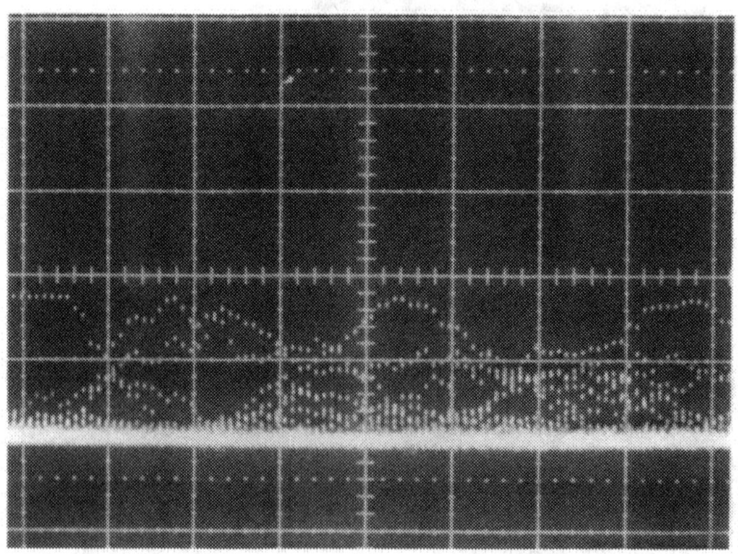

Fig. D.8 *Output from correlation receiver for time dispersive channel*

The spectrum of the direct sequence modulated test signal consists at all times of 2 × 63 spectral components equally spaced by Δf across the bandwidth with a $\sin^2 x/x^2$ relative power envelope, as shown in Fig. D.7. The instantaneous phases of each spectral component make up the varying sequence waveform. Because of the ion density gradient across the ionospheric layer, the point of reflection of the different spectral components will vary, giving rise to dispersion. The larger the bandwidth, the larger is the variation in the point of reflection between the outermost spectral components. A high correlation of the diffracted waves representing the spread spectrum components will, however, have a similar effect as the presence of specular component. Thus applying this model to the test results reported indicates a lack of any steady specular component but a sufficient correlation in the diffracted components up to 80 kHz bandwidths. Above this the 160 kHz bandwidth performs poorly, possibly owing to an incoherent breakup of the signal spectrum. The rare situation of no significant drop in performance for the 160 kHz bandwidth could be due to a steady specular component or a temporary increase in the correlation of the diffracted wave.

Figs. D.2*a*, D.3*a* and D.4*a* show the output from the passive correlator only for a time equivalent to a few sequence periods. In this case the multipaths seem fairly steady. However, an increase in the time span shows time fading multipaths, as illustrated in Fig. D.8.

Reference

1 SKAUG, R.: 'Experiments with spread spectrum modulation on radiowaves reflected from the ionosphere', *Arch. Elektron. & Üebertragungstech.*, 1981, **35** (4), pp. 151–155

Index